ENVIRONMENTAL HEALTH - PHYSICAL, CHEMICAL AND BIOLOGICAL FACTORS

CRUDE OILS

PRODUCTION, ENVIRONMENTAL IMPACTS AND GLOBAL MARKET CHALLENGES

ENVIRONMENTAL HEALTH - PHYSICAL, CHEMICAL AND BIOLOGICAL FACTORS

Additional books in this series can be found on Nova's website under the Series tab.

Additional e-books in this series can be found on Nova's website under the e-book tab.

ENVIRONMENTAL HEALTH - PHYSICAL,
CHEMICAL AND BIOLOGICAL FACTORS

CRUDE OILS

PRODUCTION, ENVIRONMENTAL IMPACTS AND GLOBAL MARKET CHALLENGES

CLAIRE VALENTI
EDITOR

New York

For permission to use material from this book please contact us:
Telephone 631-231-7269; Fax 631-231-8175
Web Site: http://www.novapublishers.com

Additional color graphics may be available in the e-book version of this book.

Library of Congress Cataloging-in-Publication Data

ISBN: 978-1-63117-950-1

Published by Nova Science Publishers, Inc. † New York

CONTENTS

PREFACE

The primacy of crude oil as an energy source has provided a platform for social-economic development in most oil producing and consumer nations. Food, housing, transportation, investment services and inflationary trend is largely dependent on the volatility of crude oil prices. The anticipated shift in crude oil demand premised on the United States shale exploitation and the economic pandemonium generated in the Organization of Petroleum Exporting Countries (OPEC) has re-enacted crude oil pricing as the barometer for global economic trend. Crude oil exploitation, production, transportation and utilisation processes, however adversely impact on marine and coastal habitat. Pollution and infections attributable to toxic substances from crude oil and its derivatives can have long and short term health implications on animals and humans. Crude oil dependence is also contributory to global warming and its complications from climate change. This book examines the environmental and global market impact of crude oil production, analyses attempts to curb inherent challenges through energy efficiency and the development of renewable alternative energy sources.

Chapter 1 - Stable water-in-oil emulsions, often formed after oil spills, contribute to the difficulties of cleanup due to their persistence and high viscosity. The objectives of the present study were to determine the fate and effects on grass shrimp (*Palaemonetes pugio*) and blue crabs (*Callinectes sapidus*) of such emulsions after they enter estuaries. To achieve this objective, non-emulsified oil and stable emulsions, formed from Kuwait crude oil, were added to estuarine mesocosms, followed by exposure of grass shrimp (*Palaemonetes pugio*) to treated sediments. The polycyclic aromatic hydrocarbon (PAHs) concentrations in the mesocosm with emulsified oil decreased from 284 to 7 μg/g sediment in 56 days, while in the mesocosm with non-emulsified oil the PAHs decreased from 271 to 0.2 μg/g sediment over this same time period. Reproduction parameters (ovary development, embryo production) of grass shrimp were affected as result of exposure to sediments with emulsified oil, including no embryo production (Day 14 sediments) and reduced embryo production (Day 36 sediments). In contrast, grass shrimp reproduction parameters were not affected after exposure to sediments with the same concentration of non-emulsified oil. It is suggested that the persistence of emulsified oil explains the observed effects. Exposure of grass shrimp embryos to pore water from emulsified oil sediments resulted in significantly more DNA strand breaks and reduced embryo hatching rates compared to reference controls or to sediments with non-emulsified oil. In addition to work with oiled sediments, a histological study was conducted on blue crabs fed food containing emulsified oil. The most notable effect was distended hemocytes with large amounts of glycoproteins in the hepatopancreas. It is speculated that crabs with these distended hemocytes are less able to deal with invading microbes, since crab hemocytes are an important part of the crab's immune system. This study suggests that the entrance of water-in-oil emulsions into estuaries can effect grass shrimp reproduction. Procedures that inhibit emulsion formation, thus preventing emulsified oil from entering estuaries, should be considered after oil spills.

Chapter 2 - The formation of wax crystals in engine oils at low temperatures due to poor solubility of paraffins profoundly affects the oil rheology. In fact, gelation occurs when growing paraffin crystals interlock and form a volume-spanning crystal network which entrains the remaining liquid oil among the crystals. These macroscopic structures may cause yield stress. Currently, the rapidly increasing use of mineral base oils having considerably higher paraffinic content than past solvent-refined base oils calls for careful identification of gelation tendency because gelation can produce a serious failure in the mechanism of engines called air-binding.

Nowadays, there are two different methods commonly used to evaluate gelation: the Mini Rotary Viscometer test (MRV), ASTM D 4684 and the Scanning Brookfield test (SBT), ASTM D 5133. The test ASTM D 4684 is incorporated in the SAE J300 viscosity classification and both test methods are part of ILSAC GF-5 and GF-6 standards. Obviously, the simultaneous use of two test methods (which could not be correlated) to evaluate the same phenomena (gelation) would have the risk of reducing the range of lubricants available to the formulator.

This work carries out an experimental study of both tests by means of the use of a carefully selected set of lubricant oils. Preliminary results show the utility of the test ASTM D 5133 for Newtonian lubricants because this test permits the use of the Scanning Brookfield Viscometer as an alternative to the Mini Rotari for testing Newtonian liquids in order to check the standard SAE J300. The authors observed by means of the SBT the breakdown of the wax network of the gelation process of a lubricant oil subjected to shearing. In this case the SBT provides wrong information. Therefore, the appearance of a normal gelation curve is a necessary requirement to assure the validity of the measurements provided by the Scanning Brookfield Viscometer.

Chapter 3 - The chapter considers the price relationships in crude oil futures by using stepwise regression. It follows the publication by the CFTC of three-year historical data for the disaggregated version of its Commitment of Traders report. Physical, macroeconomic and financial determinants of the price of oil are included as regressors, and their respective explanatory power is interpreted based on t-tests. The empirical results are based on ARMAX models. By using weekly data from 2006 until present, the regression analysis sheds light on the link between the CFTC 'Money Managers' (large investors) category and price movements, and suggests the presence of 'excessive speculation' during the 2008 oil price swing episode, after controlling explicitly for a wide range of determinants.

Chapter 4 - Fuel production from renewable resources has been intensified worldwide. Biofuels have the advantage of causing less environmental pollution than petroleum fuels. Brazil stands among the countries with the largest biodiesel production, 2.72 billion liters in 2012, according to recent data from the National Agency of Petroleum, Natural Gas and Biofuels (ANP). For each liter of biodiesel produced 100 ml of crude glycerin is generated, therefore Brazil alone had a production of 272 million liters of crude glycerin. As a global consequence, we currently have a glycerin production above the market demand, which requires researches in order to ensure new ways of using this co-product. In addition to enabling an extra profit in the supply chain, defining new alternatives to using glycerin is important to minimize an environmental obstacle, as to date there is still no well-defined law on how the discharge of excess glycerin must occur. Poultry industry is a highly developed economic activity in Brazil and in other countries such as the U.S.A., China and EU countries. Another important fact is that these countries also highlighted the global biodiesel production. Thus, poultry production has been recognized as an interesting alternative to increase the demand of glycerin. Several studies have evaluated the use of crude or processed glycerin in animal feed and showed that glycerin can be considered a good source of dietary energy for poultry, also representing an opportunity that fits the need of producing quality meat with environmental responsibility. During the discussion in this chapter the results of the most significant studies on glycerin use in poultry nutrition published in high impact journals were considered. To date, it is not possible yet to define a single glycerin inclusion level in poultry diet that is properly applicable to all the different situations in animal production. Overall, the compilation of the work contemplated in this review suggests that the safe rate of glycerin inclusion is 5% for broilers, 7.5% for laying hens and from 4 to 8% for quails.

Chapter 5 - The primacy of crude oil as an energy source has provided a platform for social-economic development in most oil producing and consumer nations. Food, housing, transportation, investment services and inflationary trend are largely dependent on the volatility of crude oil prices. The anticipated shift in crude oil demand premised on the United States shale exploitation and the economic pandemonium generated in the Organization of Petroleum Exporting Countries (OPEC) has re-enacted crude oil pricing as the barometer for global economic trend. Crude oil exploitation, production, transportation and utilisation processes, however adversely impact on marine and coastal habitat. Pollution and infections attributable to toxic substances from crude oil and its derivatives can have long and short term health implications on animals and humans. Crude oil dependence is also contributory to global warming and its complications from climate change. This paper examines the environmental and global market impact of crude oil production, analyses attempts to curb inherent challenges through energy efficiency and the development of renewable alternative energy sources. It concludes that the political instability, insecurity, corruption and economic adversity confronting nation states, particularly oil producing developing states, has compromised their zeal to tackle crude oil induced market and environmental problems – stockpiling crude.

Chapter 6 - Due to increasingly strict vehicle fuel economy mandates over the past two decades, fuel economy improvement continues to be a focal point in all aspects of engine and vehicle engine and operation. This includes engine oil formulation, whose fuel economy improvement potential can be estimated in the interval from 1% to 4%, depending on the chosen baseline. In response, all major global regions have established standard engine oil fuel economy tests, whether through industry groups (such as ILSAC, API, or ACEA), or through individual OEMs. In addition to the standardized fuel economy tests that oils are required to pass to meet specifications, bench tests have historically been used to screen and assess the fuel economy performance of these oils. It is generally accepted for engine oils that fuel economy improvement is influenced by reductions in kinematic viscosity, high shear viscosity, boundary friction, thin-film friction and pressure-viscosity coefficient (or traction coefficient in elastohydrodynamic lubrication). These measurements are relatively quick and easy to obtain compared to the more sophisticated and expensive engine performance tests. Therefore, optimization of bench tests can be considered as a fascinating challenge for petroleum industry. Nevertheless, in order to evaluate the fuel economy performance of lubricant oils by means of bench tests it is required the previous design of a precise predictive model. Currently, the Spanish company REPSOL S.A. in cooperation with the Research Group of Physics and Chemistry of Linares (University of Jaén-Spain) is developing a new model to characterize the CEC L-54-96 standard engine fuel economy test which is part of the standards ACEA A1/B1, A5/B5, C1, C2, C3 and C4. Preliminary results and reflections are shown in this commentary.

In: Crude Oils
Editor: Claire Valenti

ISBN: 978-1-63117-950-1

Chapter 1

EMULSIONS PRODUCED AFTER OIL SPILLS: THEIR FATE IN ESTUARIES AND EFFECTS ON THE GRASS SHRIMP, *PALAEMONETES PUGIO* AND BLUE CRAB, *CALLINECTES SAPIDUS*

***Richard F. Lee*[1*], *Keith Maruya*[2], *Ulrich Warttinger*[1], *Karrie Bulski*[1] *and Anna N. Walker*[3]**
[1]Skidaway Institute of Oceanography, University of Georgia, Savannah, GA, US
[2]Southern California Coastal Water Research Project Authority, Costa Mesa, CA, US
[3]Department of Pathology, Mercer University School of Medicine, Macon, GA, US

ABSTRACT

Stable water-in-oil emulsions, often formed after oil spills, contribute to the difficulties of cleanup due to their persistence and high viscosity. The objectives of the present study were to determine the fate and effects on grass shrimp (*Palaemonetes pugio*) and blue crabs (*Callinectes sapidus*) of such emulsions after they enter estuaries. To achieve this objective, non-emulsified oil and stable emulsions, formed from Kuwait crude oil, were added to estuarine mesocosms, followed by exposure of grass shrimp (*Palaemonetes pugio*) to treated sediments. The polycyclic aromatic hydrocarbon (PAHs) concentrations in the mesocosm with emulsified oil decreased from 284 to 7 μg/g sediment in 56 days, while in the mesocosm with non-emulsified oil the PAHs decreased from 271 to 0.2 μg/g sediment over this same time period. Reproduction parameters (ovary development, embryo production) of grass shrimp were affected as result of exposure to sediments with emulsified oil, including no embryo production (Day 14 sediments) and reduced embryo production (Day 36 sediments). In contrast, grass shrimp reproduction parameters were not affected after exposure to sediments with the same concentration of non-emulsified oil. It is suggested that the persistence of emulsified oil explains the observed effects. Exposure of grass shrimp embryos to pore water from emulsified oil sediments resulted in significantly more DNA strand breaks and reduced embryo hatching rates compared to reference controls or to sediments with non-emulsified oil. In addition to work with oiled sediments, a histological study was conducted on blue crabs fed food containing emulsified oil. The most notable effect was distended hemocytes with large amounts of glycoproteins in the hepatopancreas. It is speculated that crabs with these distended hemocytes are less able to deal with invading microbes, since crab hemocytes are an important part of the crab's immune system. This study suggests that the entrance of water-in-oil emulsions into estuaries can effect grass shrimp reproduction. Procedures that inhibit emulsion formation, thus preventing emulsified oil from entering estuaries, should be considered after oil spills.

INTRODUCTION

One of the consequences of oil spills in coastal waters can be the formation of stable water-in-oil emulsions with mixing energy provided by wind and tides (Daling et al., 2003; Fingas et al., 2001; Lee, 1999; Lunel et al., 1996). Stable water-in-oil emulsions are characterized by their persistence, high water content, high viscosity, small water droplets and higher density than the original oil (Brandvik and Daling, 1991; Fingas et al., 1994). The high viscosity results in a semi-solid or gel which contributes to the persistence of the emulsified oil, as well as making cleanup more difficult.

Oil, stranded after entering estuaries, can be toxic to a variety of marine invertebrates, including crustaceans, polychaetes and mollusks (Glemarec and Hussenot 1982; Jackson et al., 1989; Lee et al., 1981; McGuinness, 1990; National Research Council, 1985; Peterson, 2000; Sanders et al., 1980). However, few of these studies have compared the effects of emulsified with non-emulsified oil. For sediment exposure we used grass shrimp which can comprise 50% or more of the pelagic macrofauna biomass in estuarine tidal creeks and are an important food for many estuarine fish (Reinsel et al., 2001; Scott et al., 1992). The present study reports on the fate and effects of water-in-oil emulsions and non-emulsfied oil in estuarine mesocosms. Changes in sediment PAH concentrations over 85 days and the effects of these oiled sediments on the reproduction parameters (ovary development, embryo production), embryo hatching rates and DNA strand breaks of grass shrimp (*Palaemonetes pugio*) were determined. In addition to exposure of grass shrimp to oiled sediments, a histological study was conducted on blue crabs (*Callinectes sapidus*) fed food containing emulsified oil.

MATERIALS AND METHODS

A) Preparation of Water-in-Oil Emulsions

Water-in-oil emulsions were prepared by rotating 500ml cylindrical separatory funnels containing 300ml of estuarine water and 30ml of Kuwait crude oil. The rotating apparatus run at 30 rpm for up to 24 hrs was similar to the one described by Hokstad et al. (1993). Water content of the emulsions were noted at 0.5, 1, 2, 4, 6, 8, 12 and 24 hours and the data used to calculate an uptake $t_{½}$ defined by Hokstad et al. (1993) as the time needed to pick up half the maximum water content. Water droplet sizes in the emulsion were determined with an Olympus microscope equipped with an image analyzer system and software (Image Pro Plus). Both white light and blue fluorescence light were used in determining water droplet diameters.

B) Addition of Emulsions to Mesocosm

The mesocosms used were 1m deep with a volume of $300m^3$ and each contained approximately 1000kg of estuarine sediment (collected from the Skidaway River estuary, Savannah, GA). A water-in-oil emulsion (360 ml of emulsion) made from Kuwait crude oil as described above was added to a 30 x 30cm section in the center of the mesocosm. To a second mesocosm non-emulsified Kuwait crude oil (360ml) was added. A third mesocosm, where no oil was added, served as a reference control.

C) Analysis of Polycyclic Aromatic Hydrocarbons (PAH) in Sediment

Sediment cores were taken from the control mesocosm and the oil treated mesocosms at days 1, 14, 28, 42, 56, 70 and 84. Procedures for the analysis of sediment sediment cores for PAHs were similar to those described earlier (Maruya et al., 1997). Sediment samples (10g) from each core were freeze dried, extracted with methylene chloride and extracts passed through a silica gel column. The PAH fractions from the silica gel column were analyzed with a Hewlet Packard 6890 Series Plus gas chromatograph coupled to a 5973 mass spectrometer. Twenty three individual PAH analytes were quantified with a detection limit of 10ng/g for each analyte. Except for naphthalene, recovery of spiked PAH standards ranged from 89 to 95%, while naphthalene recovery ranged from 55 to 67%. Procedural blanks, NIST sediments, spiked matrices and replicate samples were analyzed using the same procedures outlined above. For PAH spiked sediments, recoveries were between 90-100%.

D) Grass Shrimp Reproduction Assay

For this assay, approximately 1 kg of sediment was taken from each mesocosm (control, treatment with emulsified Kuwait crude oil and treatment with non-emulsified Kuwait crude) at each time period (Day 14, 36 and 56). Sediments from each time period were subdivided into 3 parts and one part added to each of 3 aquaria. Estuarine water (40L to each aquarium) and grass shrimp (20 shrimp to each aquarium), collected from a nearby estuary (Skidaway Island, GA) was combined with crude oil contaminated or control sediment. The salinity of the water was 25-27 ppt and the temperature in the aquaria was maintained at 27°C by aquarium heaters. The grass shrimp were fed frozen *Artemia* sp. and kept under a 12 h light/12 h dark regime. The number of females in each aquaria was determined and reproduction parameters for these females were followed. Every 5 days the following parameters were determined in each aquarium: (a) mortality; (b) number of females with mature ovaries; (c) number of females with attached embryos. Grass shrimp were exposed to sediments from each time period for 50 days with no water change. Forty-eight embryos (stage 8) from an egg bearing female from each aquarium were transferred to a 48 well polystyrene plate with each well containing 1.1ml of estuarine water and 1 embryo. Culture plates were kept in the dark at 27°C and embryo hatching rates were determined by daily examination of plates under a dissecting microscope. Hatching was generally completed within 48h after transfer to culture plates. Data is reported as mean ± standard deviation (n=3)

E) Grass Shrimp Embryos with Sediment Pore Waters

Pore waters were obtained by centrifugation (1500 x g) from oiled and non-oiled mesocosm sediments. Sediment collections were conducted on day 1, 7, 14, 28, 49 and 87. Grass shrimp embryos (stage 7) were removed with forceps from egg bearing females. Description of different embryo stages of grass shrimp has been described in Winston et al. (2004). One grass shrimp embryo was placed in each well of a 48 well polystyrene tissue culture plate. Each well contained 1.1ml of diluted sediment pore water [1:10 diluted with filtered (0.25 μm filter) estuarine water]. Plates were kept at 27°C in an incubator. Hatching rates [percentage of embryos which hatched into the zoea stage (stage 12)] were determined by daily examination of embryos in the plates under a dissecting microscope.

F) Comet Assay on Grass shrimp Embryos Exposed to Sediment Pore Water

Twenty embryos (stage 7) from the same female used for the hatching test described above were added to a beaker containing 20ml of diluted pore water (1:10 diluted with estuarine water) from each mesocosm and incubated at 27°C for 24 hours. The embryos were removed from the beaker after 24 hours and used for the comet assay. The procedures for the comet assay were a modification of the procedures described by Singh et al. (1988) and Steinert et al. (1998). Embryos were placed in a microcentrifuge tube on ice, followed by addition of 1ml of 4°C filtered estuarine water to each tube. Embryos were homogenized in a 1ml ground glass tissue homogenizer and left to settle at 4°C for 5 min. During this settling period individual cells remained in suspension while intact tissue pieces and fragments sank to the bottom. A portion of the supernatant (800μl) was transferred to a second microcentrifuge tube, followed by centrifugation at 14,000 x g for 2 min at 4°C. After removal of the supernatant, the pelleted cells were resuspended in 70μl of 0.65% low-melting point agarose in Kenny's salt solution (0.4M NaCl, 9mM KCl, 0.7 mM K_2HPO_4, 2 mM $NaHCO_3$), cooled to ≈25°C after melting. This suspension was pippetted onto a slide coated with 1% agarose and a coverslip applied. The agarose was allowed to harden on ice-cold trays for 3min. Slides were transferred to lysis buffer (2.5M NaCl, 0.1M EDTA, 0.01M TRIS-HCl, 10% dimethyl sulfoxide, 1% Triton X100, pH10). After incubation overnight at 4°C, slides were rinsed with water (4°C). After this water rinse, slides were placed in unwinding buffer (0.2 M NaOH, 1 mM EDTA, pH>13.5, 4°C) for 15 min. Electrophoresis of slides in the unwinding buffer was carried out at 300mA and 25V for 20 min. At the end of the electrophoresis run slides were neutralized by rinsing slides in 0.4M TRIS-HCl (pH 7.5) and gels were fixed by rinsing in 100% ethanol (4°C) for 5 min. Once dry, cells were stained with 20μl of ethidium bromide (20μg/ml). DNA strand breaks were determined by quantitation of fluorescence (excitation at 540 nm) in the head and tail of 50 randomly chosen cells from each replicate slide using an inverted fluorescent microscope (Nikon Eclipse E400), high sensitivity CCD camera and an image analysis system (Komet 4.0 - Kinetic Imaging Ltd.). Data is reported as the mean ± SEM (n=5) reported as percent of DNA in the tail where 100 cells/animal were scored. Datasets were compared using 1-way ANOVA followed by Tukey's post-test. Statistically significant differences were expressed as $P < 0.05$ or $P < 0.01$.

G) Emulsified Oil in Food Fed to Juvenile Blue Crabs

Juvenile blue crabs (*Callinectes sapidus*) were fed food (brine shrimp) containing water-in-oil emulsions (25μg/g food). Controls were fed clean brine shrimp. At different times (3 and 7 days) the hepatopancreas was removed from crabs fixed in 10% neutral buffered formalin containing 1% zinc formalin. The fixed tissues were processed, embedded in paraffin, sectioned, stained with hematoxylin and eosin and examined by light microscopy.

RESULTS

Emulsion Formation

Stable water-in-oil emulsions with water content of 95% and mean water droplet size 3.1 ± 1.5μm were formed after mixing Kuwait crude oil with estuarine water (Figures 1 and 2). These oil emulsions were added to mesocosm sediments. In a separate experiment a small amount of emulsified oil and non-emulsified oil (approximately 200g) was added to a nearby estuary. After seven days the highly viscous emulsified oil remained firmly adsorbed to the sediment, while most of the non-emulsified oil was washed away by incoming tides and only an oily stain remained on the sediment (Figure 3).

Changes in PAH Concentrations over Time

Changes in total and individual PAH concentrations in the oil treated mesocosms are given in Figures 4 & 5. The data shown are only for the upper core sections (1-3cm) since elevated PAH concentrations were not observed in the oiled sediments for the 3-6 cm sections (data not presented). Total PAH concentrations decreased with time in both the emulsified and non-emulsified oiled sediments, but the decrease was much more pronounced in sediment with non-emulsified oil (Figure 4 & 5). Oiled sediments showed a typical petrogenic PAH profile with high concentrations of lower molecular weight alkalated PAHs, e.g. methylnaphthalenes and phenanthrenes (Figure 5). The unoiled control sediment showed a low background concentration of total PAHs (0.1 μg/g sediment) composed primarily of high molecular weight pyrogenic type PAHs (4-5 ringed nonalkylated PAHs; data not shown).

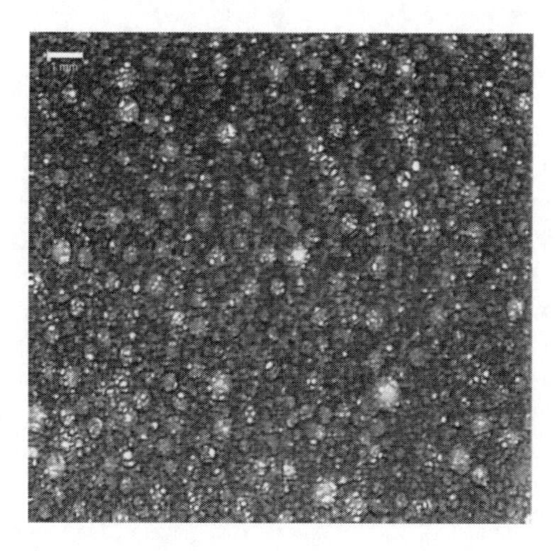

Figure 1. Photomicrograph of water-in-oil emulsion formed by Kuwait crude oil with estuarine water (taken 12 hours after mixing). Note small size of water droplets (1 to 10 microns) which is characteristic of stable emulsions.

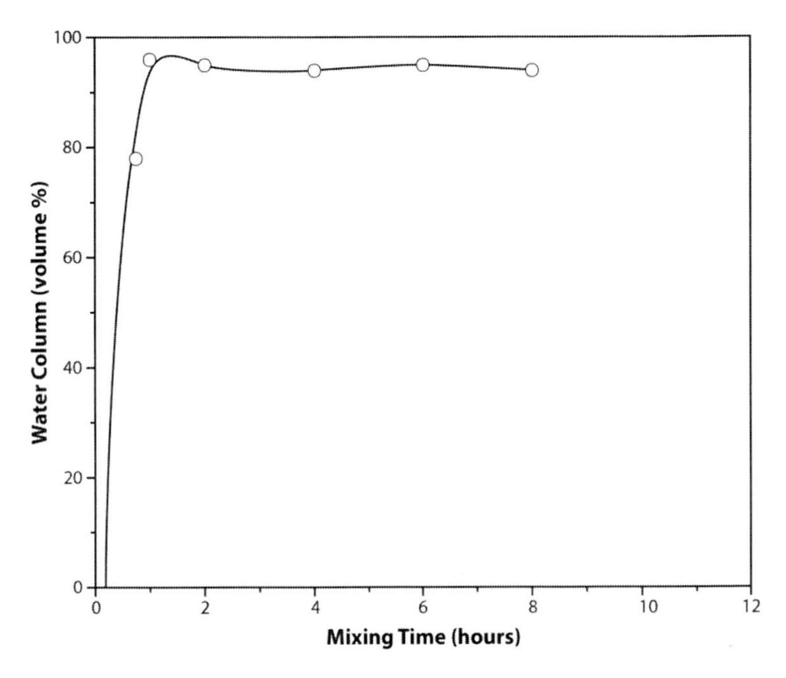

Figure 2. Take up of water during the formation of an oil-in-water emulsion. Kuwait oil (30ml) was mixed with estuarine water (300ml) with water content measured at the different times.

Figure 3. Emulsified Kuwait crude oil (200g; upper left) and non-emulsified Kuwait (200g; bottom right) added to estuarine sediment. Photograph taken 4 days after addition of the oils to the estuarine sediment.

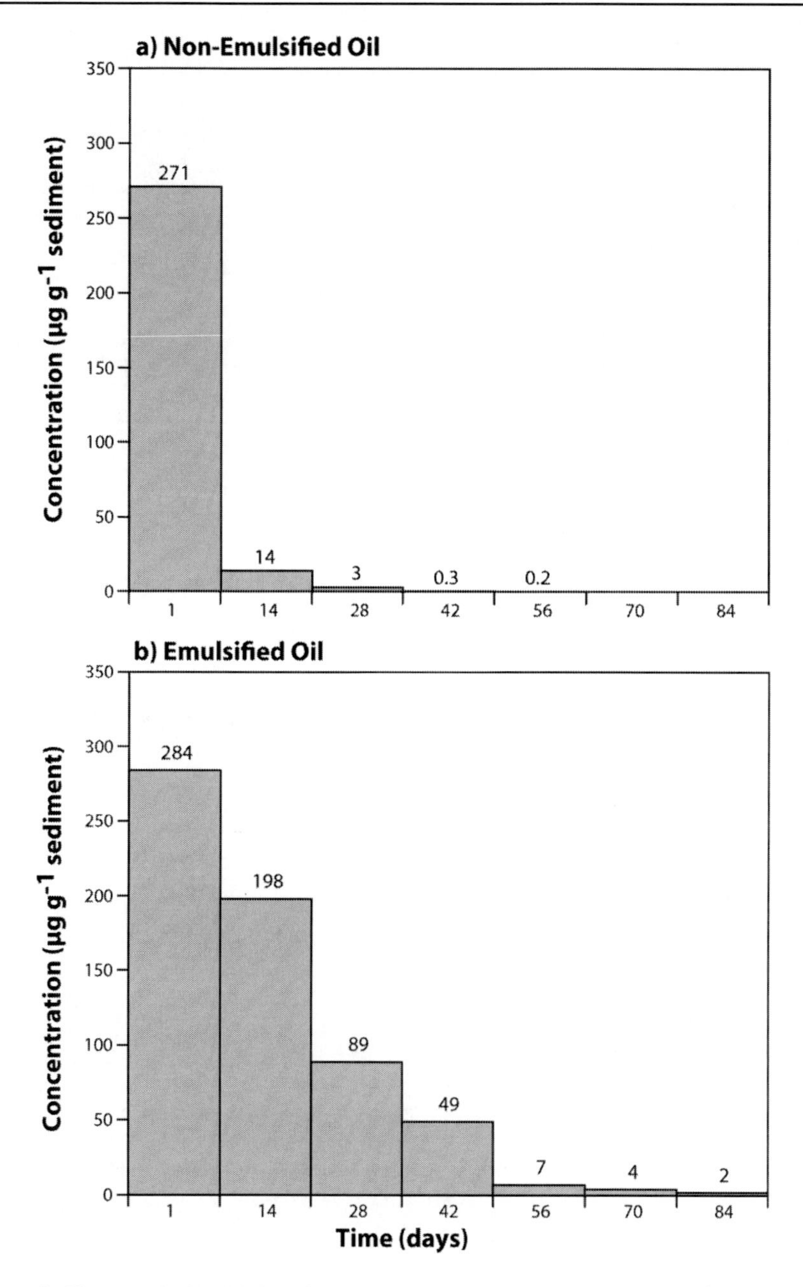

Figure 4. Changes during 84 days in total PAHs concentrations from sediments with emulsified and non-emulsified oils.

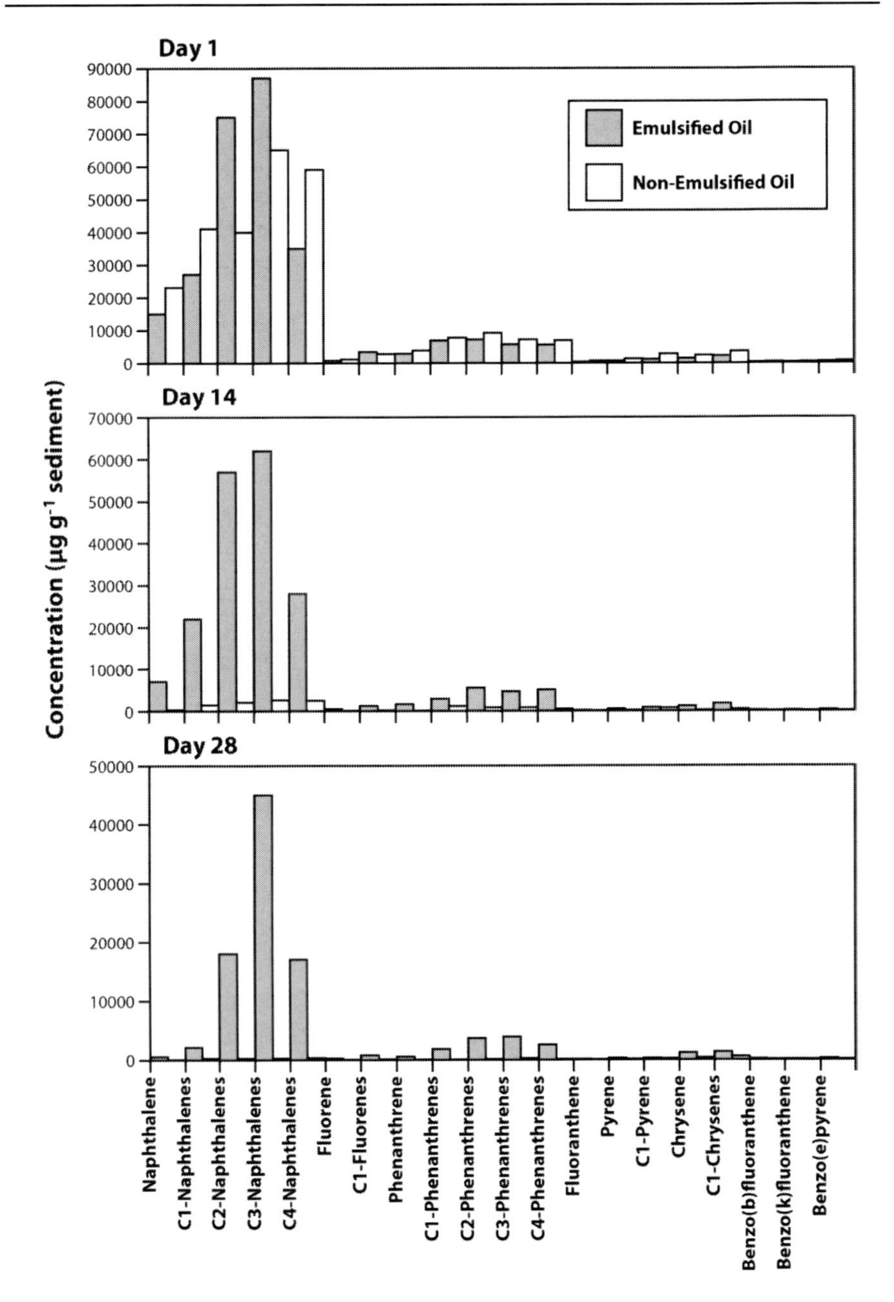

Figure 5. Changes during 28 days in the concentration of individual PAHs in sediments with emulsified and non-emulsified oils.

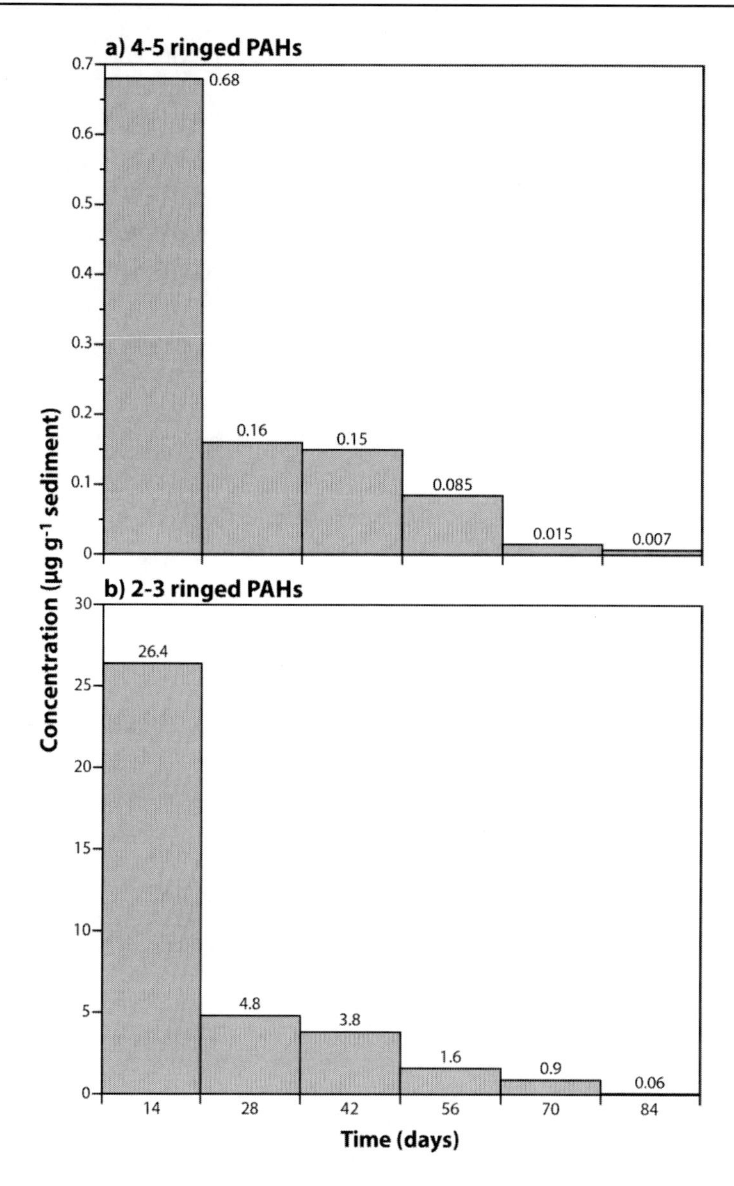

Figure 6. Changes during 84 days in the concentrations of 2-3 ringed PAHs and 4-5 ringed PAHs from sediments with emulsified oil.

Total sediment PAH concentrations at Day 1 in control, emulsified oil and non-emulsified oil treatments were 0.1, 271 and 284µg/g sediment, respectively, while on Day 14, PAH concentrations were 0.5, 198 and 14µg/g sediment, respectively (Figure 4). By Day 28 the non-emulsified oiled sediment had low total PAHs (3μg/g sediment) with little evidence of petrogenic PAHs, while the emulsified oiled sediment still contained relatively high total PAH concentrations (89μg/g sediment), predominantly petrogenic PAHs. At day 84 the total PAH in sediment with emulsified oil was 2μg/g sediment, but there was still evidence of petrogenic PAHs. Figure 6 compares the changes in the concentrations of the 2-3 ringed PAHs with 4-5-ringed PAHs in sediments with emulsified oil. The 2-3 ringed PAHs were initially the major PAH groups but rapidly decreased while the 4-5 ringed PAHs also decreased over time but were the most persistent PAH group.

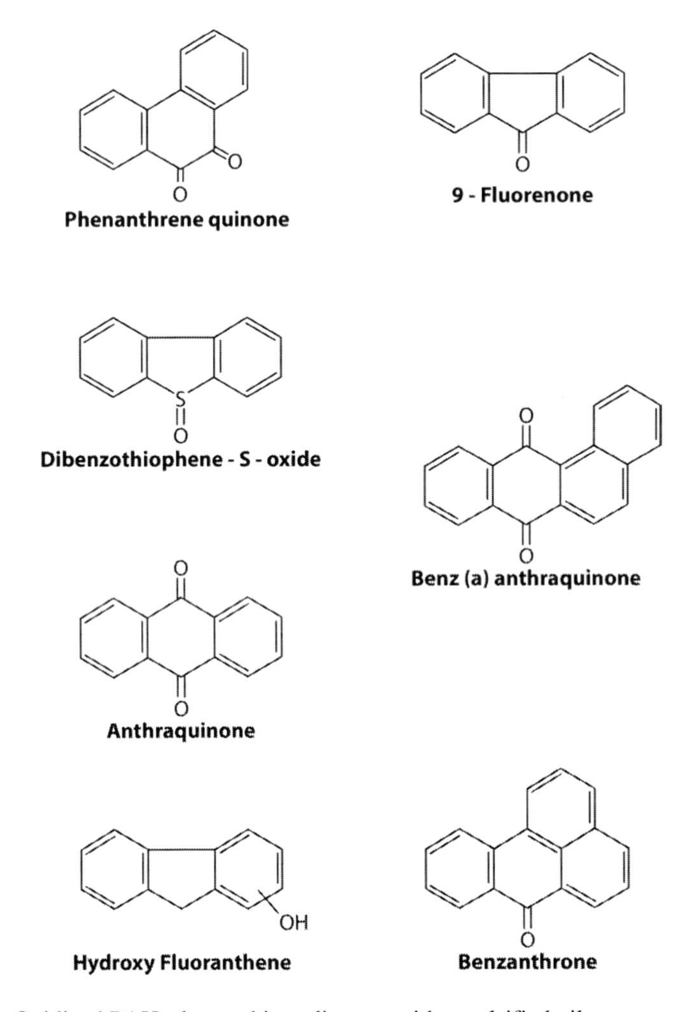

Figure 7. Oxidized PAHs detected in sediments with emulsified oil.

A number of oxidized PAHs were detected in the oiled sediments at Days 14 and 28, including phenanthrene quinone, dibenzothiophene-S-oxide, 9-fluorenone, anthraquinone, benzanthrone, benz(a)anthraquinone and hydroxyfluoranthene (Figure 7) The presence of these oxidized PAHs indicates that a variety of oxidation processes, e.g. photo-oxidation, acted on the stranded oil. These compounds were not detected in the fresh crude oil (data not presented).

Mortality of Grass Shrimp Exposed to Oiled Sediments

Preliminary laboratory toxicity work showed high mortality of grass shrimp within 48 hours when 100 g of sediments were treated with 200μl of emulsified Kuwait crude oil. However there was no grass shrimp mortality when 50 and 100μl of emulsified oil were added to 100g of sediment. To avoid mortality, based on these initial toxicity experiments, the amount of oil added to the three mesocosms were as follows: (1) 360ml of emulsified oil to one mesocosm (~1000kg of sediment); (2) 360ml of non-emulsified oil to a second mesocosm; (3) no oil was added to the third mesocosm.

Effects of Oiled (Emulsified and Non-Emulsified Oils) Sediments on Reproduction and Embryo Development of Grass Shrimp

The effects (reproduction parameters and embryo hatching rates) as a result of exposure of grass shrimp to oiled and non-oiled sediment from the different time periods are presented in Table 1. The 14 day emulsified oiled sediments affected reproduction (lack of embryo production, reduction in the number of females with mature ovaries). Shrimp exposed to the 36 day emulsified oiled sediments showed significantly reduced embryo production compared with controls, but there was no significant effects on embryo hatching rates. No effects were observed for reproduction parameters or embryo hatching rates as a result of exposure to 58 day old emulsified oil sediment. Exposure of shrimp to non-emulsified oil for all time periods (14, 36 and 58 days) showed no significant effects compared to controls on reproduction parameters or embryo hatching rates.

Table 1. Assessment of reproduction, embryo production, embryo hatching rates and DNA strand breaks in grass shrimp exposed to oiled (emulsified oil and non-emulsified oil) and reference sediments[1]

Time (days)	Reproduction (% females forming mature ovaries) Mean ± S.D.(n=3)	Embryo Production (% females producing embryos) Mean ± S.D. (n=3)	Hatching (% embryos hatching into zoea stage) Mean ± S.D.(n=3)	DNA Strand Breaks Comet Assay (% DNA in tail) Mean ± S.E.(n=5)
Emulsified Oil				
14	33 ± 8*	0*	-	-
36	68 ±10	20 ± 8*	88 ± 6	15.7 ± 2.7*
58	77 ± 8	27 ± 15*	84 ± 6	11.2 ± 3.3
Non-Emulsified Oil				
14	82 ± 10	43± 15	86 ± 9	10.9 ± 2.9
36	78 ± 8	47 ± 16	92 ± 4	8.2 ± 3.3
58	77 ± 13	43 ± 8	87 ± 8	7.9 ± 3.0
Reference Sediment				
14	83 ± 10	44 ± 16	90 ± 6	8.9 ± 2.8
36	79 ± 8	52 ± 14	88 ± 5	9.4 ± 4.1
58	82 ± 10	52 ± 12	90 ± 8	8.8 ± 3.5

[1] The time refers to the number of days the sediments (oiled or unoiled) were in the mesocosm before transfer to aquaria. See Methods for details of these experiments.

* Denotes significant difference from reference sediment at $P<0.05$ (ANOVA, F-test).

Tests with Sediment Pore Water

The effects of pore water from emulsified oiled sediments, non-emulsified oiled sediments or reference sediments on embryo DNA (strand breaks – comet assay) and embryo hatching rates are summarized in Table 2. The comet assay detects DNA strand breaks and alkali labile sites by measuring the migration of DNA from immobilized nuclear DNA (Figure 8). Sediment pore waters from emulsified oiled sediments caused significantly more DNA strand breaks and reduced embryo hatching rates than pore water from reference or non-emulsified oiled sediments.

Table 2. Assessment of embryo hatching rates and DNA strand breaks in grass shrimp embryos (stage 8) exposed to pore water from oiled (emulsified oil and non-emulsified oil) and reference sediments. Time refers to time after oil was added to a particular mesocosm

Sample Time (days)	Hatching (% embryos hatching into zoea stage) Mean ± S.D. (n=3)	DNA Strand Breaks Comet Assay (% DNA in tail) Mean ± S.E. (n=5)
Pore Water from Sediment with Emulsified Oil		
14	14 ± 4*	41.3 ± 9.1*
36	27 ± 4*	21.3 ± 3.2*
58	55 ± 18	17.4 ± 4.2*
80	89 ± 11	11.2 ±2.8
Pore Water from Sediment with Non-Emulsified Oil		
14	85 ± 6	9.7 ± 3.3
36	91 ± 7	12.2 ± 2.6
58	89 ± 8	8.2 ± 3.4
80	88 ± 11	11.4 ± 2.7
Pore Water from Reference Sediment		
14	91 ± 6	7.8 ± 2.0
36	90 ±5	5.9 ± 2.4
58	85 ± 8	6.2 ± 3.1
80	88 ± 10	4.9 ± 3.4

* Denotes significant difference from reference pore water at $P<0.05$ (ANOVA, F-test).

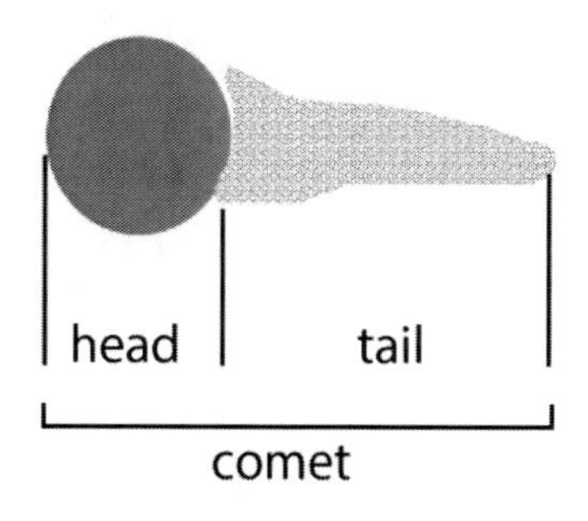

Figure 8. Diagram of typical comet from grass shrimp embryos showing distribution of DNA in tail and head.

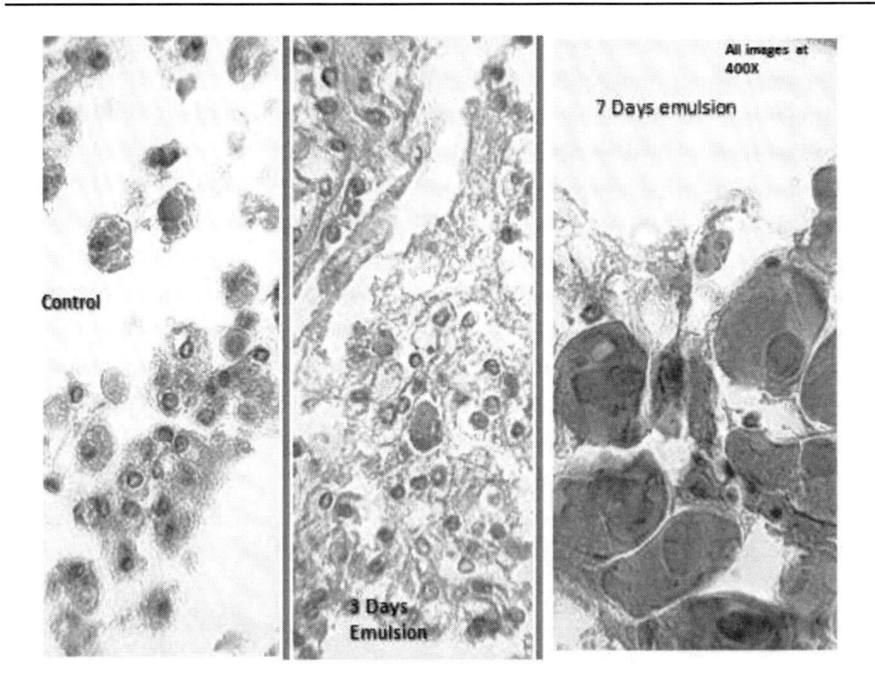

Figure 9. Photomicrograph of hepatopancreas of control crab and crab fed food containing emulsified oil (25μg/g food) for 3 and 7 days.

Blue Crab Studies

The hepatopancreas of juvenile blue crabs fed food containing emulsified oil (25 μg of emulsified oil/g of food) had distended hemocytes filled with eosinophilic materials (Figure 9). While distended hemocytes were observed on day 3, it was particularly pronounced on day 7. Results of the various stains (Gomori Trichrome: phosphotungstic acid-hematoxylin: Periodic acid-Schiff stain after treatment with distase) suggested the eosinophilic material is a glycoprotein.

Discussion

PAH concentrations in sediments treated with non-emulsified oil decreased rapidly and were close to background in 14 days, whereas petrogenic PAHs were still detected 84 days after sediments were treated with emulsified oil. The initial PAH concentrations (198 μg/g) were relatively low compared to PAH concentrations (6000 to 100,000 μg PAH/g sediment) that have been reported in sediments after oil spills (Lee and Page, 1997).

In the present study sediments containing emulsified oil (271 μg/g sediment) affected the reproduction parameters and embryo hatching rates of the grass shrimp, *Palaemonetes pugio*. These reproduction parameters were not affected by similar concentrations of non-emulsified oil. The persistence of PAHs in emulsified oiled sediments would explain the continued effects of these weathered sediments on the reproduction parameters of grass shrimp. During an actual spill, the emulsified oil would not enter the estuary at a single time point but would likely enter repeatedly at different times. It can be assumed that stranded emulsified oil, where PAH concentrations are very high (~ 10, 000 μg/g) would show continued effects on grass shrimp reproduction for many months after the spill. Petroleum compounds in water can be toxic to grass shrimp, both adults and their embryos (Fisher and Foss, 1993; Mearns et al., 1995; Rayburn et al., 1996; Shelton et al., 1999; Unger et al., 2007). PAHs in sediment can affect various reproduction parameters in grass shrimp, including ovary development, embryo production and embryo hatching (Lee et al., 2008; Oberdörster et al., 2000; Wirth et al., 1998). The results of the present study suggest that emulsified oil effects reproduction parameters of grass shrimp at lower concentrations than does non-emulsified oil.

A very rapid (1-2 days) and inexpensive assay, the sediment pore water assay, was used to assess genotoxocity and embryotoxicity in oiled and unoiled sediments (Table 2). This assay assessed DNA strand breaks and embryo hatching rates after grass shrimp embryos were exposed to sediment pore water. Earlier studies showed that PAHs in sediment could cause DNA strand breaks in embryos and adults of grass shrimp (Lee and Steinert, 2003; Lee et al., 2004, 2008). The results with pore water assays were quite comparable to the results obtained when juvenile grass shrimp were exposed to oiled sediments (compare reproduction parameters/embryo hatching rates in Table 1 to DNA strand breaks/embryo hatching rates in Table 2). Thus, it appears that an inexpensive and rapid pore water assay could be used to rapidly assess the genotoxicity of sediments over a broad area after an oil spill. This would allow a search for "hot spots" in a coastal area where a spill has occurred and also allow rapid monitoring of post-spill conditions.

In addition to the effects of emulsified oil on grass shrimp, a histology study was conducted on juvenile blue crabs fed food containing emulsified oil. The most notable effect was distended hemocytes with large amounts of glycoproteins in the hepatopancreas. Crab hemocytes are an important part of the crab immune system by participating in the encapsulation, phagocytosis and melanization of invading microbes (Cerenius et al., 2010; Jiravanichpaisal et al., 2006; Shields and Overstreet, 2007). We speculate that crabs with such distended hemocytes, as a result of exposure to emulsified oil, are less able to deal with invading microbes. Exposure to a number of anthropogenic chemicals has been shown to affect the immune systems of invertebrates (Galloway and Depledge, 2001). For example, lobsters exposed to various pesticides can have reduced hemocyte phagocytosis (De Guise et al. 2004, 2005).

CONCLUSION

Stable water-in-oil emulsions, often formed after oil spills, can persist in estuarine sediments. This study found that the rate of PAH decrease was much slower in sediments with emulsified oil compared with sediment with non-emulsified oil. This higher PAH concentration in sediments with emulsified oil explains why such sediments affect reproduction parameters, embryo hatching rates and DNA strand breaks of grass shrimp more than sediments with non-emulsified oil. If emulsified oil enters an estuary, it can be expected to affect various macrofauna, such as grass shrimp and blue crabs, which play an important role in estuarine ecology.

ACKNOWLEDGMENTS

This publication was made possible by EPA grant number 83518401. Its contents are solely the responsibility of the grantee and do not necessarily represent the official views of the EPA. Further, the EPA does not endorse the purchase of any commercial products or services mentioned in the publication.

REFERENCES

Brandvik, P.J. & Daling, P.S. *W/o-emulsion formation and w/o-emulsions stability testing – an extended study with eight oil types.* DIWO Report No. 10. Trondheim, Norway; IKU SINTEF Group; 1991.

Cerenius, L., Jiravanichpaisal, P., Liu, H-P. & Söderhäll, I. (2010). Crustacean immunity. In: Söderhäll, K., editor. *Invertebrate Immunity.* New York: Springer Science+Business Media 2010: pp. 239-259.

Daling, P.S., Moldestad, M. Ø., Johansen, Ø., Lewis, A. & Rødak, J.(2003). Norwegian testing of emulsion properties at sea – The importance of oil type and release conditions. *Spill Science & Technology Bulletin 8:* 123-136.

De Guise, S. Maratea, J. & Perkins, C. 2004. Malathion immunotoxicity in the American lobster (homarus americanus) upon experimental exposure. *Aquat. Toxicol.* 66:419-425.

DeGuise, S., Maratea, J., Chang, E.S. & Perkins, C. 2005. Resmethrin immunotoxicity and endocrine disrupting effects in the American lobster (*Homarus americanus*) upon experimental exposure. *J. Shellfish Res.* 24:781-786.

Fingas, M.F., Fieldhouse, M., Mullin, J. Studies of water-in-oil emulsions and techniques to measure emulsion treating agents. In: *Proceedings of the Arctic Marine Oilspill Program Technical Seminar.* Ottawa, Canada: Environment Canada; 1994; pp. 213-244.

Fingas, M.F., Fieldhouse, B., Lane, J. & Mullin, J.V. What causes the formation of water-in-oil emulsions? In: *Proceedings of the 2001 International Oil Spill Conference.* Washington, DC: American Petroleum Institute 2002: pp. 109-114.

Fisher, W. & Foss, S.S. (1993). A simple test for toxicity of number 2 fuel oil and oil dispersants to embryos of grass shrimp, *Palaemonetes pugio. Mar. Pollut. Bull.* 26:385-391.

Galloway, T.S. & Depledge, M.H. (2001). Immunotoxicity in invertebrates: measurement and ecotoxicological relevance. *Ecotoxicology* 10:5-23.

Glemarec, M. and Hussenot, E. (1982). A three-year ecological survey in Benoit and Wrac'h Abers following the *Amoco Cadiz* oil spill. *Netherlands J Sea Res* 16:483-490.

Hokstad, J.N., Daling, P.S., Lewis, A.S. & Shrom-Kristiansen, T. 1993. Methodology for testing water-in-oil emulsions and demulsifers. Description of laboratory procedures. In: Walker AH, Ducey DL, Gould JR, Orvik SB, eds, *Formation and Breaking of Water-In-Oil Emulsions: Workshop Proceedings.* MSRC Technical Report Series 93-018. Marine Spill Response Corp., Washington, DC, pp 239-353.

Jackson, J.B.C., Cubit, J.D., Keller, B.D., Batista, V., Burns, K., Caffey, H.M., Caldwell RL, Garrity SD, Getter CD, Gonzalea C, Guzman HM, Kaufman KW., Knap, A.H., Levings, S.C., Marshall, M.J., Steger, R., Thompson,

R.C. & Weil, E. (1989). Ecological effects of a major oil spill on Panamanian coastal marine communities. *Science* 243:37-44.

Jiravanichpaisal, P., Lee, B.L. & Söderhäll, K. 2006. Cell-mediated immunity in arthropods: hematopoiesis, coagulation, melanization and opsonization. *Immunobiology* 211:213-236.

Lee, R.F. (1999). Agents which promote and stabilize water-in-oil emulsions. *Spill Science & Technology* 5:117-126.

Lee, R.F. & Page, D.S. (1997). Petroleum hydrocarbons and their effects in subtidal regions after major oil spills. *Mar. Pollut. Bull.* 34:928-940.

Lee, R.F. & Steinert, S. (2003). Use of the single cell gel electrophoresis/comet assay for detecting DNA damage in aquatic (marine and freshwater) animals. *Mutation Res.* 544:43-64.

Lee, R.F., Dornseif, F., Gonsulin, F., Tenore, K. & Hanson, R. (1981). Fate and effects of a heavy fuel oil spill on a Georgia salt marsh. *Mar Environ. Res* 5:125-143.

Lee, R.F., Maruya, K.A. & Bulski, K. (2004). Exposure of grass shrimp tosediments receiving highway runoff. Effects on reproduction and DNA. *Mar. Environ. Res.* 58:713-717.

Lee, R.F., Niencheski, L.F.H. & Brinkley, K. (2008). DNA strand breaks in grass shrimp embryos exposed to highway runoff sediments and sediments with coal fly ash. *Mar. Environ. Res.* 66:164-165.

Lunel, T., Revin, J., Bailey, N., Halliwell, C. & Davis, L. A successful at sea response to the *Sea Empress* spill. In: *Proceedings of the 19th Arctic Marine Oilspill Program Technical Seminar,* Ottawa, Canada, Environment Canada; 1996; pp. 1499-1520.

Maruya, K.A., Loganathan, B.G., Kannan, K., McCumber, S. & Lee, R.F. (1997). Organic and organometallic compounds in estuarine sediments in the Gulf of Mexico (1993-1994). *Estuaries* 20:700-709.

McGuinness, K.A. (1990). Effects of oil spills on macro-invertebrates of saltmarshes and marine forests in Botany Bay, New South Wales, Australia. *J Exp Mar Biol Ecol* 142:121-135.

Mearns, A., Doe, K., Fisher, W., Hoff, R., Lee, K., Siron, R., Mueller, C. & Venosa, A. Toxicity trends during an oil spill bioremediation experiment on a sandy shoreline in Delaware, USA. In: *Proceedings of the Arctic Marine Oilspill Technical Seminar.* Ottawa, Canada: Environment Canada, 1995; pp. 1133-1145.

National Research Council. *Oil in the Sea: Inputs, Fates, and Effects. National Research Council*, National Academy Press, Washington, DC.(1985).

Oberdörster, E., Brouwer, M, Hoexum-Brouwer, T., Manning, S. & McLachlan, J.A. (2000). Long-term pyrene exposure of grss shrimp, *Palaemonetes pugio*, affects molting and reproduction of exposed males and offspring of exposed females. *Environmental Health Perspectives* 108:641-646.

Peterson, C.H. (2000). The *Exxon Valdez* oil spill in Alaska: acute, indirect and chronic effects on the ecosystem. *Adv. Mar. Biol.* 39:3-84.

Rayburn, J.R., Glas, P.S., Fossie, S.S. & Fisher, W.S. (1996). Characterization of grass shrimp (*Palaemonetes pugio*) embryo toxicity tests using the water soluble fraction of number 2 fuel oil. *Mar. Pollut. Bull.* 32:860-866.

Reinsel, K.A, , Glas, P.S., Rayburn, J.R., Pritchard, M.K. & Fisher, W.S. (2001). Effects of food availability on survival, growth and reproduction of the grass shrimp *Palaemonetes pugio*: a laboratory study. *Mar Ecol Prog Ser* 220:231-239.

Sanders, H.L, Grassle, J.P, Hampson, G.R., Morse, L.S., Garner-Price, S. & Jones, C.C. (1980). Anatomy of an oil spill: long-term effects from the grounding of the barge *Florida* off West Falmouth, Massachusetts. *J Mar Res* 38:265-380.

Scott, G.I., Fulton, M.H., Crosby, M.C., Key, P.B. & Daugomah, J.W. (1982). *Agriculture insecticide runoff effects of estuarine organisms: correlating laboratory and field toxicity tests, ecophysiology bioassays and ecotoxicological biomonitoring.* US Environmental Protection Agency, Gulf Breeze Environmental Research Laboratory, Gulf Breeze, FL.

Shelton, M.E., Chapman, P.J., Foss, S.S. & Fisher, W.S. (1999). Degradation of weathered oil by mixed marine bacteria and the toxicity of accumulated water-soluble material to two marine crustacean. *Arch. Environ. Contam. Toxicol.* 36:13-20.

Shields, J.D. & Overstreet, R.M. 2007. *Diseases, parasites, and other symbionts.* In: Kennedy, V l& Cronin, L.E. (Eds.), The Blue Crab *Callinectes sapidus*. College Park, MD: Maryland Sea Grant; 2007; pp. 299-417.

Singh, N.P., McCoy, M.T., Tice, R.R. & Schneider, E.L. (1988). A simple technique for the quantification of low levels of DNA damage in individual cells. *Exp. Cell Res.* 175:184-191.

Steinert, S.A., Streib-Montee, R., Leather, J.M. & Chadwick, D.B. (1998). DNA damage in mussels at sites in San Diego Bay. *Mutat. Res.* 399:65-85.

Winston, G.W., Lemaire, D.G.E. & Lee, R.F. (2004). Antioxidants and total oxyradical scavenging capacity during grass shrimp, *Palaemonetes pugio*, embryogenesis. *Comp. Biochem. Physiol.* 139C:281-288.

Wirth, E.P., Fulton, M.H., Chandler, G.T., Kay, P.B. & Scott, G.I. (1998). Toxicity of sediment associated PAHs to the estuarine crustaceans, *Palaemonetes pugio* and *Amphiascus tenuiremis. Bull. Environ. Contam. Toxicol.* 61:617-644.

In: Crude Oils
Editor: Claire Valenti
ISBN: 978-1-63117-950-1

Chapter 2

IMPROVEMENTS IN THE TESTING OF LUBRICANT OILS AT LOW TEMPERATURES CARRIED OUT BY INDUSTRIAL LABORATORIES

José-Alberto Maroto-Centeno[1,*], Tomás Pérez-Gutiérrez[2,†] and Manuel Quesada-Pérez[1,‡]

[1]Group of Physics and Chemistry of Linares, Department of Physics, EPS of Linares, University of Jaén, Jaén, Spain

[2]Technology Unit., Repsol, Móstoles, Madrid, Spain

ABSTRACT

The formation of wax crystals in engine oils at low temperatures due to poor solubility of paraffins profoundly affects the oil rheology. In fact, gelation occurs when growing paraffin crystals interlock and form a volume-spanning crystal network which entrains the remaining liquid oil among the crystals. These macroscopic structures may cause yield stress.

* Corresponding author. Group of Physics and Chemistry of Linares, Department of Physics, EPS of Linares, University of Jaén, C/ Alfonso X el Sabio, 28, 23700 Linares, Jaén, Spain. Tel.: +34 953 648 553. E-mail: jamaroto@ujaen.es.

† E-mail: tperezg@repsol.com.

‡ E-mail: mquesada@ujaen.es.

Currently, the rapidly increasing use of mineral base oils having considerably higher paraffinic content than past solvent-refined base oils calls for careful identification of gelation tendency because gelation can produce a serious failure in the mechanism of engines called air-binding.

Nowadays, there are two different methods commonly used to evaluate gelation: the Mini Rotary Viscometer test (MRV), ASTM D 4684 and the Scanning Brookfield test (SBT), ASTM D 5133. The test ASTM D 4684 is incorporated in the SAE J300 viscosity classification and both test methods are part of ILSAC GF-5 and GF-6 standards. Obviously, the simultaneous use of two test methods (which could not be correlated) to evaluate the same phenomena (gelation) would have the risk of reducing the range of lubricants available to the formulator.

This work carries out an experimental study of both tests by means of the use of a carefully selected set of lubricant oils. Preliminary results show the utility of the test ASTM D 5133 for Newtonian lubricants because this test permits the use of the Scanning Brookfield Viscometer as an alternative to the Mini Rotari for testing Newtonian liquids in order to check the standard SAE J300. We observed by means of the SBT the breakdown of the wax network of the gelation process of a lubricant oil subjected to shearing. In this case the SBT provides wrong information. Therefore, the appearance of a normal gelation curve is a necessary requirement to assure the validity of the measurements provided by the Scanning Brookfield Viscometer.

INTRODUCTION

Satisfactory operation of engines at low temperature is dependent on the ability of the oil to provide good pumpability oil flow. Pumpability is a very complex process because it includes behaviour of the lubricant in different parts of the engine [1]. In fact, an engine can be divided into three zones in which different shear rates and shear stresses are encountered by the lubricant [2]. Zone A corresponds to the flow of oil from the sump to the oil screen under gravity. Very low shear stresses (0.3 to 3.0 Pa) and shear rates (0.01 to 0.1 s^{-1}) are found in this zone [2]. If the oil flow is too low because the oil is gelled, air-binding will take place. Zone B corresponds to the flow of the oil form the screen into the pump. It is considered a higher shear zone with shear stresses in the range of 100 to 250 Pa at a viscosity of 10 Pa·s [3-5]. Zone C, from the pump to the moving surfaces, is possibly the one where the highest shear stresses are encountered. Rough calculation based on a pressure of 138 kPa and a gallery diameter of 6.35 mm give yield stress values at the gallery of the wall of approximately 2000 to 4000 Pa [6]. Zone C is by far the most complex to model and the most dependent on engine design.

Under ambient conditions (t > O°C), unused lubricating oils typically exhibit Newtonian rheology at shear rates less than 100 s^{-1}; therefore, designing the lubricated system for adequate hydrodynamic lubrication is straightforward. However, when the temperature falls well below the saturation temperature for the n-paraffin-like components of the oil, the nucleation and growth of crystalline structures can lead to highly non-Newtonian rheology. Under these conditions, oil supply can be impeded by high viscosity and/or gelation. For these reasons, the Society of Automotive Engineers incorporated a measure of low-temperature viscosity and gelation in the Engine Oil Viscosity Classification standard SAE J300, the Mini Rotary Viscometer test (ASTM D 4684). Other methods, such as the Scanning Brookfield test (ASTM D 5133), have also been proposed for inclusion in this standard. Both test methods are currently part of ILSAC GF-5 and GF-6 standards.

The Mini Rotary Viscometer test (MRV), ASTM D 4684 [7], cools lubricants to a final test temperature following a slow cooling cycle labeled TP-1 [8] and then measures a yield stress (if any) and a low shear-rate viscosity. Yield stress is a measure of the tendency of the oil to resist flow, for example, due to gelation resulting from structures formed from wax crystals. Low shear-rate viscosity is a measure of the pumpability of the oil. Both conditions, for example a large yield stress or a high viscosity, may lead to engine failure because of inadequate lubricant supply. The former is often associated with a failure mechanism called air-binding, that is, the uptake of air into the oil pump rather than lubricant; the later is commonly associated with a mechanism of flow-limited failure, that is the inability of the pump to provide enough lubricant to the engine in a timely manner [9]. MRV limits in SAE J300 have historically been based upon cold-room testing of gasoline engines and supported by a consortium of engine builders, oil and additive companies.

Another test proposed for measuring pumpability relevant low temperature properties of oils is the Scanning Brookfield test (SBT), ASTM D 5133 [10]. Its use and application to reference oils has been extensively described by Selby [11-13]. The SBT measures viscosity continuously at constant shear rate as the oil is cooled at a constant rate of 1 °C/h from -5 to -45 °C. Selby has reported that some mineral oil lubricants exhibit an abrupt increase in viscosity over a narrow temperature range, and suggested that the strength of the viscosity increase positively correlated with the severity of gelation or gel strength; he established the term "gelation index". A gelation index limit requires imposing a maximum acceptable rate for viscosity increase with decreasing temperature. It also requires demonstrating that gelation index correlate to both a physically intuitive measure such as yield stress and to phenomena relevant to engine operation at low temperatures. Selby [12] showed that increasing SBT gelation index correlated with increasing MRV TP-1 yield stress for Pumpability Reference Oils (PRO) used in an industry-wide study in the mid-1980's. ILSAC GF-5 and GF-6 currently impose a maximum gelation index of 12.

Although there are many issues that should be dealt with in deciding on the veracity of using a gelation index measure as part of a standard such as SAE J300 [14], three technical issues, in particular, are relevant. First, the physical phenomena associated with wax formation and its effects on low temperature rheology are generally not well understood. For example, it is not obvious that there is or should be a correlation between the dependence of a continuous shear rate viscosity on temperature and the onset of the structure formation that is required to support a stress, that is, a yield stress. Second, there is legitimate debate over whether or not the gelation index measurement allows identifying potential problem oils that are not identified by existing test methods such as the MRV TP-1 test. Third, several studies on larger data sets of mineral oil lubricants show no correlation between gelation index and yield stress, in apparent contradiction to Selby's work on the limited number of Pumpability Reference Oils [15].

This work carries out an experimental study of both tests by means of the use of a carefully selected set of lubricant oils. On the other hand, a deeper knowledge of the tests commonly used for the evaluation of rheological behavior of lubricant oils at low temperatures could permit the optimization of the measurement protocols used in Industrial Laboratories.

MATERIALS

The results showed in this paper are a little part of a wide research project carried out by Repsol in cooperation with the Research Group of Physics and Chemistry of Linares (University of Jaén, Spain). The experimental measurements were carried out in the Repsol advanced laboratory of lubricants, which is placed in the Repsol Centre at Móstoles (Madrid, Spain). The rheology of 8 lubricants oils has been evaluated by means of the tests ASTM D 5133 and ASTM D 4684. These 8 lubricant oils were selected between a group of commercial base oils commonly used by Repsol. In some cases a standard pour point depressant additive (PPD) was used in order to change the rheological properties of the oils at low temperatures.

THEORETICAL

A) Main Features of the Gelation Curve

The Scanning Broofield test (SBT) provides the gelation curve of a engine oil. The McCoull-Walther-Wright (MWW) is an empirical equation that has been widely used for the description of the variation in viscosity as a function of temperature in engine oils ([16-18]) and it takes the form:

$$Ln Ln \eta (T) = a - G Ln T \quad (1)$$

where η is the viscosity, T the absolute temperature and both a and G are constant. More specifically, the positive constant G is called gelation value in the context of the Scanning Brookfield Technique [19]. The MWW equation successfully reproduces the behaviour of engine oils in Newtonian regime. In fact, this equation forms the basis for the ASTM viscosity-chart, one of the most popularly methods used to determine the viscosity of oils [20]. With regard to the behaviour of engine oils at low temperatures, Selby et al. have proved experimentally that the MWW equation fits the variation of the viscosity with temperature of an engine oil at temperatures lower than the gelation index temperature, that is, for an oil which is supposed to be undergoing a crystal growth process [21].

The MWW equation is incorporated in the software of the Scanning Brookfield Viscometer in order to draw the gelation curve and evaluate some important parameters as gelation index and gelation index temperature.

In a gelation curve the incremental derivative of the MWW equation (that is, G in Equation (1)) is plotted as a function of the temperature. Figure 1 shows the main features of the gelation curve for a gelating oil [22]. This system behaves like a perfect Newtonian oil in a wide interval of temperatures in which G takes a constant value (zone I). It exhibits a quick (pulse-shaped) change of G values at lower temperatures (zone II). This behaviour has been traditionally attributed (without further proof) to the appearance of a nucleation process.

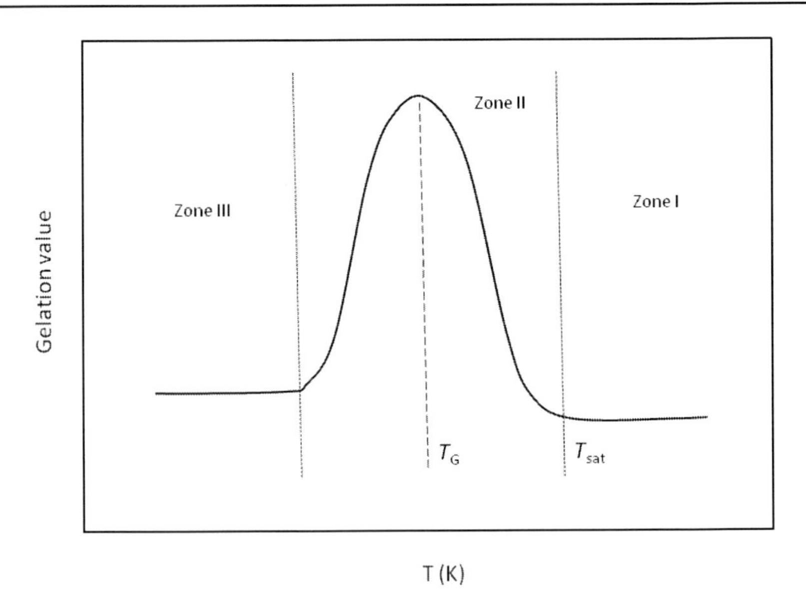

Figure 1. Gelation curve.

Therefore, T_{sat} is commonly known as saturation temperature and it is the onset temperature for nucleation [23]. The maximum gelation value in zone II is called the gelation index and the temperature for the gelation index is commonly known gelation index temperature, T_G. On the other hand, a typical gelating oil usually takes a new constant value of G for a further decrease in temperature (zone III).

Recently, Maroto el al. have carried out a theoretical work that sheds light on the physical meaning of the gelation index, that is, the main parameter involved in the test ASTM D 5133 [24,25]. More specifically, it is proved that the pulse-shaped region of the gelation curve can be explained as the relief of supersaturation of the solution, that is, a nucleation process that finishes when the saturation ratio takes a value equal to 1.

B) Comparison between the Test ASTM D 5133 and the Test ASTM D 4684

The aim of this area of petroleum industry is to ensure lubricant pumping performance at low temperatures through rheological measurements using the Mini Rotary Viscometer and Scanning Brookfield tests. Often these tests provide conflicting information, because lubricant formulations must be optimized to meet requirements of both tests [15]. It must be underlined that both ASTM methods show evident advantages and disadvantages. ASTM D 4684 method does not permit the evaluation of the temperature at which gelation, if any, takes place; also, this method reports a poor reproducibility in the evaluation of yield stress, which causes important differences between measurements carried out for different laboratories [26]. With regard to ASTM D 5133 there are still important doubts about the utility associated to the gelation index, which is the main parameter provided by this test.

Recently [15] it has been found no correlation between gelation index and yield stress (parameter with a physical meaning closely related to a gelation phenomena), in apparent contradiction to Selby work based on the study of a number of Reference Oils (PRO) used in an industry wide study in the mid-1980's [12]. These results suggest that the process occurring in the region characterized by the gelation index, which is the onset of wax crystallization through nucleation and growth of wax crystallites, is not directly related to the occurrence of gelation in the oil.This study has come to the conclusion that the gelation index from ASTM D 1533 is not an adequate relative measure of gelation in mineral oils. Nowadays, the test ASTM D 4684 is incorporated in the SAE J300 viscosity classification and both test methods are part of ILSAC GF-5 and GF-6 standards. Obviously, the simultaneous use of two test methods (which could not be correlated) to evaluate the same phenomena (gelation) would have the risk of reducing the range of lubricants available to the formulator.

From a mathematical point of view, the following functionality can be written for the viscosity of a lubricant oil:

$$\eta(T) = f[T, \left(\frac{dT}{dt}\right), \left(\frac{d\gamma}{dt}\right)_C, \left(\frac{d\gamma}{dt}\right)_M, COMP] \tag{2}$$

where T the absolute temperature, (dT/dt) is the cooling rate applied during cooling, $(d\gamma/dt)_c$ is the shear rate applied during cooling, $(d\gamma/dt)_m$ is the shear rate applied during the measurements and *COMP* is the composition of the oil.

Equation (2) can be used to analyze the influence of the operating conditions of the Scanning Brookfield Viscometer and the Mini Rotary in the viscosity measurements carried out by these instruments or, in other words, this equation can be useful to answer the following fundamental question: Why are different in many cases the viscosity measurements carried out by both instruments at the same temperature T on the same oil with composition

COMP?. In order to answer this question we must take into account that the rest of parameters in equation (2) are rather different for both instruments. On the one hand, in the case of the Scanning Brookfield Viscometer $(dT/dt) = 1°C/h$ and $(d\gamma/dt)_c = (d\gamma/dt)_m = 0.2\ s^{-1}$; on the other hand, in the case of the Mini Rotary (dT/dt) is characterized by the TP-1 cycle, $(d\gamma/dt)_c = 0$ and $0.4\ s^{-1} \leq (d\gamma/dt)_m \leq 15\ s^{-1}$. Probably, the most remarkable difference between both techniques is the continuous shearing of the oil during cooling carried out by the Scanning ($(d\gamma/dt)_c \neq 0$) and the quiescent cooling of the oil carried out by the Mini Rotary ($(d\gamma/dt)_c = 0$). Table 1 summarizes these data.

The effect of the continuous shearing on yield stress exhibited by gelating oils have been analysed by Venkatesan et al. [27]. They come to the conclusion that the presence of a continuous shearing results in two competing effects: the tendency to aggregate the precipitating crystals and the tendency to break down the growing crystals. As the gelation shear rate increases, the breakdown tendency increases. Hence, the maximum yield stress occurs when the shearing is just enough to achieve maximum size of crystals, without being high enough to break down the structure. In our opinion, these conclusions can be easily extrapolated to the viscosity exhibited by a gelating oil. In fact, Selby has suggested [19] that the shearing conditions imposed by the Scanning Brookfield Viscometer promotes the growing process of paraffin crystals, which impedes the flow of the remaining oil and causes an increase of viscosity greater than in quiescent conditions.

With regard to the role played by the shear rate applied during the measurements, Webber has shown that gelating oils exhibit a shear thinning behaviour [23]. From this point of view, we can only talk about "apparent viscosity".

Ideally, the best way to research the feasible correlation between the measured reported by both tests (ASTM D 5133 and ASTM D 4684) is by analysing independently the role played by each of three parameters showed in Equation (2) (that is, (dT/dt), $(d\gamma/dt)_c$ and $(d\gamma/dt)_m$). With regard to dT/dt, that is, the cooling cycle, it is important to underline that the software of the Mini Rotari permits programming additional cooling cycles (different to those saved in its data base) and, particularly, the cooling cycle used in the Scanning Brookfield Viscometer. On the other hand, and is spite of the directives of the standard ASTM D 5133, which recommends continuous stirring from -5°C up to the final temperature of the test, the Scanning Brookfield Viscometer permits the cooling of the oil in quiescent conditions up to the desirable temperature in which this instrument starts the measurements.

Table 1. Parameters that characterize SBT and MRV

	dT/dt	$(d\gamma/dt)_c$	$(d\gamma/dt)_m$
SBT	1 °C/h	0.2 s^{-1}	0.2 s^{-1}
MRV	TP-1 cycle	0	0.4 s^{-1} - 15 s^{-1}

In fact, Kinker et al. carried out measurements with the Scanning Brookfield Viscometer in quiescent conditions in order to evaluate the role played by the continuous shearing on the oil rheology [1]. As general conclusion, the only parameter that can not be modified in both instruments is $(d\gamma/dt)_m$.

RESULTS AND DISCUSSION

Figure 2 shows both the viscosity versus the absolute temperature and the gelation value versus the temperature for the oil A2 (5W/20). This figure demonstrates that in absence of a nucleation process (that is, in absence of zone II in Figure 1) the viscosity data provided by SBT coincide with the viscosity data provided by the Mini Rotary. The explanation for this behaviour is very simple: in absence of paraffin crystallization the oil behaves like a Newtonian liquid and, therefore, the viscosity is only a function of the absolute temperature T and the composition of the oil *COMP*. Therefore, the other three key parameters that characterize the viscosity measured by both SBT and Mini Rotary ((dT/dt), $(d\gamma/dt)_c$ and $(d\gamma/dt)_m$) do not play any role for Newtonian Liquids. Similar results have been found by Selby et al [22].

In our opinion, this remarkable feature is important because the Scanning Brookfield Viscometer can be used as an alternative to the Mini Rotari for testing Newtonian liquids in order to check the standard SAE J300.

The oil E2 (5W/30) was used to research the effect of a standard pour point depressant additive (PPD) in the rheology of a lubricant oil. Figure 3 shows the viscosity provided by both the Mini Rotary and Scanning viscometers versus the Celsius temperature for the oil E2 containing proportions of PPD in the interval [0%, 0,3%]. In the light of Figure 3 is evident that both the viscosity of the oil and the tendency to exhibit yield stress provided by the MRV decrease with the increment of PPD. In fact, from the point of view of the standard SAE J300 only the oil E2 containing 0.3% pour point depressant is valid for commercial use, which evidences the utility of pour point depressants to optimize the behavior of lubricant oils at low temperatures.

Nevertheless, the results provided by the SBT are not simple to analyze. In fact, these results seem to be incoherent. For example, viscosity measured at -35ºC for the oil containing 0,03% PPD is lower than the viscosities measured at -35ºC for the oil containing 0,1% and 0,3 % PPD, respectively. On the other hand Figure 4 show both the viscosity and the gelation curve measured by means of SBT for the oil E2 containing 0,3% PPD. It is evident that the appearance of the gelation curve is abnormal due to the presence of two peaks After carrying out a deep analysis of these results we found a plausible explanation in the text of the test ASTM D 5133. Concretely, epigraph X2 entitled "Abnormal Viscosity versus Temperature Curves Indicating Instruments Problems" shows figure X2.5, very similar to Figure 4, and the following comments: "Example of data that indicates that there are no instrument problems, but that the sample has very particular low temperature rheological properties in which a structure forms in the oil, is broken and then reforms".

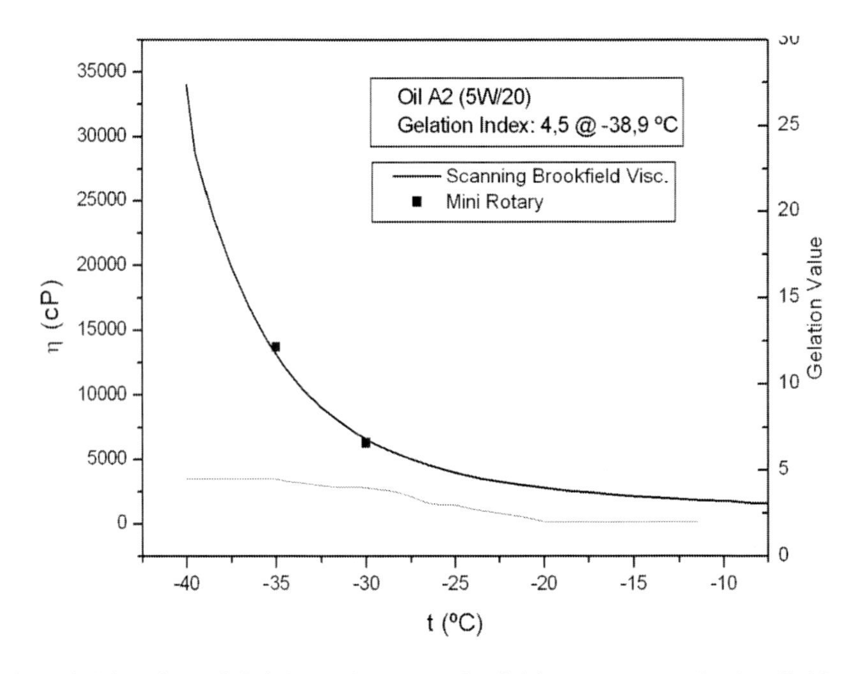

Figure 2. Viscosity and Gelation value versus the Celsius temperature for the oil A2.

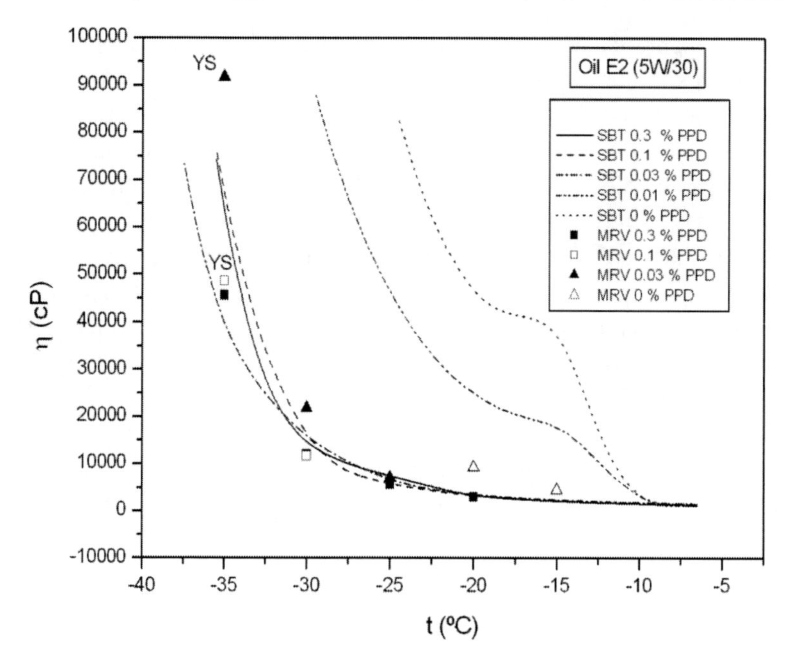

Figure 3. Viscosity versus the Celsius temperature for the oil E2 provided by MRV and SBT.

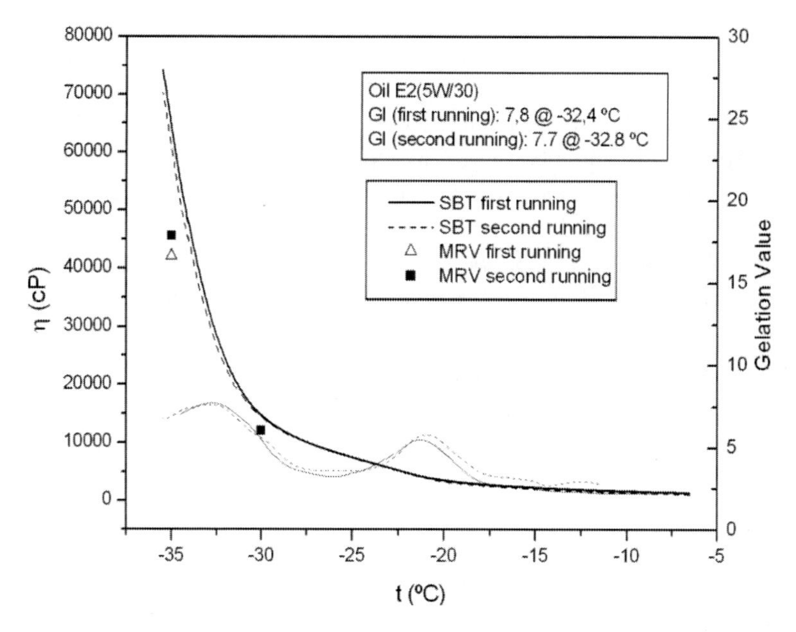

Figure 4. Gelation curve for the oil E2 containing 0,3% PPD Viscosity.

Also, Figure 4 shows the excellent repeatability in the measurements provided by the the Repsol advanced laboratory of lubricants. As conclusion we have found that the shearing of the oil brought about by the SBT may cause, in some cases, the breakdown of the wax network in a similar way that was observed by Venkatesan et al. [27]. Therefore, the appearance of a normal gelation curve (similar to that showed in Figure 1) is a necessary requirement to assure the validity of the measurements provided by the Scanning Brookfield Viscometer. From the point of view of the functionality of the viscosity of a lubricant oil given by the equation (2) is evident that the continuous shearing of the oil during cooling carried out by the scanning ($(d\gamma/dt)_c \neq 0$) is a key parameter that play a role that must be clarified in order to find hypothetical correlations between viscosity measurements provided by the MRV and the SBT. In fact, Maroto et al. have recently obtained some results that shows the slowing down of the gelation process of a lubricant oil subjected to shearing, which has been explained due to the alignment of the crystal platelets in the way of the shear and the subsequent slowing down of the crystallization process [28].

Conclusion

This work puts forward the functionality that should be used for the analysis of the real correlation between the standard tests ASTM D 5133 and ASTM D 4684. Preliminary results show the utility of the test ASTM D 5133 for Newtonian lubricants because the Scanning Brookfield Viscometer can be used as an alternative to the Mini Rotari for testing Newtonian liquids in order to check the standard SAE J300. We observed, by means of the SBT, the breakdown of the wax network of the gelation process of a lubricant oil subjected to shearing. In this case the SBT provides wrong information. Therefore, the appearance of a normal gelation curve is a necessary requirement to assure the validity of the measurements provided by the Scanning Brookfield Viscometer.

REFERENCES

[1] B. G. Kinker, J. M. Souchik and C. D. Neveu, "The Scanning Brookfield Technique: Background and Evaluation of the Low Temperature Performance of Engine Lubricants", *SAE paper* 982948, 1998.

[2] M. L. McMillan and C. K. Murphy, "The relationship of Low-Temperature Rheology to Engine Oil Pumpability", *SAE Paper* 730478, 1973.

[3] T. W. Selby, "Viscosity and the Cranking Resistance of Engine Oils at Low Temperatures", Sixth World Petroleum Congress, Frankfurt, Germany, June 19-26, 1963.

[4] F. F. Tao and W. E. Waddley, "Low Shear Viscometry and Cold Flow Mechanism – Engine Oils", *SAE Paper* 730481, 1973.

[5] R. M. Stewart and M. F. Smith, Jr., "Proposed Laboratory Methods for Predicting the Low-Temperature Pumpability Properties of Crankcase Oils", *SAE paper* 730479, 1973.

[6] T. W. Selby, "Problems in Bench Test Prediction of Engine Oil Performance at Low Temperature", *SAE Paper* 922287, 1992.

[7] "Test Method for Determination of Yield Stress an Apparent Viscosity of Engine Oils al Low Temperature", ASTM D 4684, American Society for Testing and Materials, West Conshohocken, PA.

[8] Henderson, K. O., Manning, R. E., May, C. J., and Rhodes, R. B., "New Mini-Rotary Visometer Temperature Profiles that Predict Engine Oil Pumpability", *SAE Paper* 850443, 1985.

[9] Shaub, H., "A History of ASTM Accomplishments in Low Temperature Engine Oil Rheology: 1966 – 1991", Low Temperature Lubricant Rheology Measurement and Relevance to Engine Operation, ASTM STP 1143, R. B. Rhodes, Ed. American Society for Testing and Materials, West Conshohocken, PA, 1992.

[10] "Test Method for Low Temperature, Low Shear Rate, Viscosity/ temperature Dependence of Lubricating Oils Using a Temperature Scanning Technique, ASTM D 5133, American Society for Testing and Materials, West Conshohocken, PA.

[11] Selby, T. W., "Further Considerations of Low-Temperature, Low Shear-Rate Rheology Related to Engine Oil Pumpability – Information from the Scanning Brookfield Technique", *SAE Paper* 852115, 1985.

[12] Selby, T. W., "The Use of the Scanning Brookfield Technique to Study the Critical Degree of Gelation of Lubricants at Low Temperatures". *SAE Paper* 910746, 1991.

[13] Selby, T. W., "The Scanning Brookfield Technique of Low-Temperature, Low-Shear Rheology – Its Inception, Development, and Applications", Low Temperature Lubricant Rheology Measurement and Relevance to Engine Operation, ASTM STP 1143, R.B. Rhodes, Ed., American Society for Testing and Materials, West Conshohocken, PA, 1992.

[14] McFall, D., "The Weird World of Oil Gelation", Lubes 'n' Greases, 1999, May, 14-19.

[15] Webber, R. M., George, H. F., and Covitch, M. J., "Physical Processes Associated with Low Temperature Mineral Oil Rheology: Why the Gelation Index is not Necessarily a Relative Measure of Gelation". *SAE Paper* 2000-01-1806, 2000.

[16] MacCoull, N.,"Lubrication", The Texas Company, New York, 1921, p. 85.

[17] Walther, C., Erdöl und Teer, Vol. 4, 1928, p. 510.

[18] Wright, W. A., "An Improved Viscosity-Temperature Chart for Hydrocarbons," *Journal of Materials*, Vol. 4 (1), 1969, pp. 19-27.

[19] Selby, T. W. and Miiller, G. C., "Thermal History of the Engine Oil and Its Effects on Low-Temperature Pumpability and Gelation Formation," *SAE Paper* 2008-01-2481, 2008.

[20] ASTM Standard D341 – 86: Viscosity Temperature Charts f or Liquid Petroleum Products, Annual Book of ASTM Standards, ASTM International, West Conshohocken, PA, 1986.

[21] Selby, T. W. and McGeehan, J. A., "Low Temperature Rheology of Soot-Laden, Heavy-Duty Engine Oils Using the Scanning Brookfield Technique," *SAE Technical Paper Series*, 2006-01-3352, 2001, SAE International Congress & Exposition, Toronto, Canada.

[22] Selby, T. W., "Pumpability. Past Accomplishments; Present and Future challenges, Oil flow Studies at Low Temperatures in Moder Engines," *ASTM* STP 1388, H. Shaub, Ed., ASTM International, West Conshohocken, PA, 2000.

[23] Webber, R. M., "Low Temperature Rheology of Lubricanting Mineral Oils: Effects of Cooling Rate and Wax Crystallization on Flow Properties of Base Oils," *J. Rheol.,* Vol. 43(4), 1999, pp. 911-931.

[24] Maroto-Centeno, J. A., Pérez-Gutiérrez, T., and Quesada-Pérez, M., "Advances in the Understanding of Gelation in the Framework of the Test ASTM D 5133", *J. Testing and Evaluation*, Vol. 41(2), 2012.

[25] Maroto-Centeno, J. A., Pérez-Gutiérrez, T., and Quesada-Pérez, M., "Theoretical Approach to a Better Understanding of Gelation in the

Framework of Petroleum Industry: Role Played by Different Parameters", *Mat.-wiss Werkstofftech*, Vol. 44(5), 2013.

[26] Henderson, K. O., "Pumping Viscosity by Mini-Rotary Viscometer: Critical Aspects," Low Temperature Lubricant Rheology Measurement and Relevance to Engine Operation, ASTM *STP 1143*, R.B. Rhodes, Ed., ASTM International, West Conshohocken, PA, 1992, pp. 20-32.

[27] Venkatesan, R., Nagarajan, N. R., Paso, K., Yi, Y.-B, Sastry, A. M. and Fogler, H. S., "The Strength of Paraffin Gels Formed under Static and Flow Conditions", *Chemical Engineering Sci.*, 60, 2005, pp 3587 – 3598.

[28] Maroto-Centeno, J. A., Pérez-Gutiérrez, T., and Quesada-Pérez, M., "Experimental Testing and Theoretical Characterization of an Oil Gelation Process under Shearing", *J. Petrol. Sci. Eng.*, sent for publication.

In: Crude Oils
Editor: Claire Valenti

ISBN: 978-1-63117-950-1

Chapter 3

THE 2008 OIL PRICE SWING OR THE QUEST FOR A 'SMOKING GUN'

Julien Chevallier*
IPAG Business School (IPAG Lab),
Paris, France

Abstract

The chapter considers the price relationships in crude oil futures by using stepwise regression. It follows the publication by the CFTC of three-year historical data for the disaggregated version of its Commitment of Traders report. Physical, macroeconomic and financial determinants of the price of oil are included as regressors, and their respective explanatory power is interpreted based on t-tests. The empirical results are based on ARMAX models. By using weekly data from 2006 until present, the regression analysis sheds light on the link between the CFTC 'Money Managers' (large investors) category and price movements, and suggests the presence of 'excessive speculation' during the 2008 oil price swing episode, after controlling explicitly for a wide range of determinants.

JEL Codes: C32; G12; G15; Q43
Keywords: Crude Oil Futures; Price Determinants; CFTC Commitment of Traders Report; Stepwise Regression; ARMAX Model

*julien.chevallier04@univ-paris8.fr; 184 Boulevard Saint-Germain, 75006 Paris, France; Tel: +33 (0)1 49 40 73 86; Fax: +33 (0)1 49 40 72 55.

Acknowledgments

The author wishes to thank for fruitful discussions Jean-Marie Chevalier, Michel Laffitte and the Members of the Working Group on the Volatility of Oil Prices - Frédéric Baule, Frédéric Lasserre, Ivan Odonnat, Edouard Viellefond - of which the Report Chevalier (2010) has been submitted to Christine Lagarde, French Minister of Economics, Industry and Employment on February 9, 2010. The author wishes to thank also the experts consulted during the preparation of the report at the IEA, the CFTC, the U.S. Treasury, the U.S. Department of State, the Federal Reserve, the EIA, the Congressional Research Service, the U.S. Senate, the U.S. Department of Energy, the CSIS, PFC Energy, the World Bank, the IMF, Deutsche Bank (Washington, DC USA) and the European Commission (DG MARKT, DG ECFIN, DG TREN (Brussels)): Didier Houssin, David Fyfe, Jacqueline Hamra Mesa, Robert Rosenfeld, Eric Juzenas, James Moser, Nela Richardson, Stephen Sherrod, Rafael Martinez, Gregory Kuserk, Jordon Grimm, Sandra Cvitan, Douglas Hengel, Roger Diwan, Trevor Reeve, Patricia White, Howard Gruenspecht, Bob Ryan, Robert Pirog, Rena Miller, Cory Claussen, Patrick McCarty, Mark Jickling, Carmen Difiglio, Guy Caruso, Adam Sieminski, George Kramer, Thomas Helbling, Matthew Jones, Randall Dodd, Ana Fiorella Carvajal, Paulo de Sa, Shane Streifel, Masami Kojima, Robert Bacon, Jean-Christophe Donnellier, Christophe Destais, Maxime Schenckery, Cameron Griffith, Jean-Guillaume Poulain, Roland Lhomme, Hannes Huhtaniemi, Alexandre Mathis, Peer Ritter, Joan Canton, Asa Johannesson Linden, Jan Panek, Eero Ailio, Klaus-Dietman Jacobi, Marcus Lippold, Malcolm Mcdowell, Zslot Tasnadi, Adam Szolyak. All errors and omissions remain by the author.

1. Introduction

The 2008 crude oil futures price swing has drawn once more the question of oil price volatility to governments' attention. After plummeting over 145$/barrel in July 2008, West Texas Intermediate (WTI) crude oil prices crashed to 36$ in December of the same year before bouncing back to 80$ in 2009. The volatility of the crude oil futures price, as demonstrated in 2008-2009, raises a number of questions over how the price of crude oil is determined and the complex game of interdependencies between the physical and financial markets. Such a remarkable run-up in oil prices sparked a debate over whether or not crude oil prices might be driven by factors that are inherently unrelated to oil supply and demand fundamentals. Indeed, by taking long positions on crude oil futures market and by selling them a few weeks before expiration, a speculative strategy consists in re-using these gains on a sequence of long positions (on futures contracts approaching their expiry date). When commodity prices increase, the selling price is superior to the purchase price, which yields net benefits to the investor. By systematically accumulating long positions on crude oil futures markets, without physically delivering the contracts, this "financialization" of oil markets may conduct to the formation of speculative bubbles[1], with increases of the underlying spot and futures prices (Hamilton (2009)). In this view, 'speculators' would be the main culprit for the high oil price and better regulation would be needed to curb speculation[2] and bring down the price level.

Crude oil prices are usually determined on organized futures markets (U.S. WTI and European Brent contracts) according to both physical and financial fundamentals. Physical fundamentals define the dynamic balance between supply and demand: both feature a very low short-term price elasticity, which thus creates conditions of intense volatility (Cooper (2003), Saporta et al. (2009)). Financial fundamentals go beyond the petroleum market alone and contribute to the operation of financial markets as a whole, where different types of assets, including oil, are constantly competing with each other (Medlock and Jaffe (2009)). Oil therefore creates two distinct demands in the physical and financial markets: a demand for 'physical' oil and a demand for 'paper' oil. Players in these markets may have different objectives: price hedging, taking trading positions (speculation), arbitrages over time and between products, portfolio management and risk diversification, especially for indexed funds. Some participants such as investment banks can sometimes cover all these objectives (Parsons (2009)). The complexity of interactions between the physical and the financial factors therefore restricts any unequivocal explanation of the massive oil price variations in the recent period. In previous literature, we may typically

[1]The hypothesis to be tested, the presence of a speculative bubble, is of noted interest in the existing literature. See for instance Irwin et al. (2009) on this point.

[2]Speculation may be defined in a broad sense as placing funds with the understanding that the deal entails high risk. Speculators tend to rely mainly on price changes to generate profit. Note that this definition starkly differs from 'gambling', which consists on risking money on an outcome that depends mostly on chance. Speculation differs also from 'price manipulation', which consists in deliberately misleading other investors to artificially inflate or deflate market prices.

distinguish between the defenders of physical and macroeconomic fundamentals (the demand from emerging countries, the fears of a 'peak oil', the economic crisis, the role of exchange and interest rates, see Hamilton (2008), Blanchard and Riggi (2009), Kilian (2008a, 2008b, 2009, 2010), Dvir and Rogoff (2010)), and the defenders of financial fundamentals (the upsurge in 'paper' oil, the development of new products like commodity index funds, the 'herding' behaviour of investors, the action of arbitrageurs between spot and futures markets and its limits, see Büyükşahin et al. (2008), Caballero et al. (2008), Medlock and Jaffe (2009), Parsons (2009), Hamilton (2009), Tang and Xiong (2009), Kaufmann and Ullman (2009), Cifarelli and Paladino (2010), Zagaglia (2010)).

Since September 4 2009, the U.S. Commodity Futures Trading Commission (CFTC) has published more than three years of historical disaggregated data included in the Commitment of Traders (CoT) report[3]. It released a weekly time-series on the net positions in futures of different types of agents on U.S. commodity exchanges. This data provides a breakdown of each Tuesday's open interest for markets in which 20 or more traders hold positions equal to or above the reporting levels established by the CFTC. Concerning crude oil futures markets, this data concern the NYMEX Light Sweet Crude Oil. The main interest in using disaggregated CFTC data consists in decomposing the 'commercial' and 'non-commercial' agents into four sub-categories: producers and users, swap dealers (largely banks), money managers and others. As the CFTC categorizes agents as being 'primarily' commercial or non-commercial, many uncertainties arise concerning these sub-classifications as the raw data of the CoT report is not publicly available. Nonetheless, we use the disaggregated CFTC data (namely the open interest positions held by agents' sub-categories on crude oil futures) to investigate the potential 'speculative' behaviour of large investors (hedge funds, pension funds, commodity trading advisers, etc.).

Compared to previous literature, this chapter makes three important contributions. We investigate statistically the presence of three types of crude oil price fundamentals (*physical*, *macroeconomic* and *financial*) in order to disentangle the direct influence of each variable. Our first contribution therefore lies in the systematic analysis of all types of factors that are known to affect crude oil prices in the literature. Various supply variables on crude oil markets are identified in our regression analysis at statistically significant levels, along with well-known macroeconomic indicators (U.S. GDP, Consumer Sentiment, etc.).

[3]See Sanders et al. (2004) for an overview of the CFTC CoT.

Our second contribution consists in the use of CFTC disaggregated data in order to detect the potential 'speculative' behavior of market players in crude oil futures markets. We use the net positions of 'Money Managers' as our measure of speculative positions, balanced by those of companies, investment banks and others as natural hedgers. We show that the 'Money Managers' category has a statistically significant impact. Besides, we show the statistical influence of Working's (1960) Speculative T index on crude oil price changes, which suggests the presence of 'excessive' speculative pressure during the period.

Our third contribution consists in estimating our regression model with a wide range of potential *financial* fundamentals of crude oil prices that were suggested by the existing literature. This econometric strategy allows us to define the main drivers of crude oil futures prices as being a mix of OPEC's Spare Capacity, the July-August 2008 dummy variable for refineries bottlenecks (among *physical* determinants), the U.S. GDP, the University of Michigan Consumer Sentiment (among *macroeconomic* determinants), and finally the S&P GSCI Energy Spot Price Index, the CFTC variable 'Money Managers' and the Working T index (among *financial* determinants).

To summarize our results at the outset, this chapter suggests that purely financial factors have become an independent and material driver of crude oil futures prices. Through a careful econometric analysis of the influence of each kind of fundamentals, our results bring new evidence on the role played by purely 'financial' players on the 2008 oil price swing, in addition to other causes. This empirical evidence is consistent with the view that investors' positions anticipate future changes in economic conditions, making futures markets more efficient by revealing information about future demand and supply conditions.

The remainder of the chapter is structured as follows. Section 2 provides background information on crude oil futures price developments. Section 3 develops the econometric analysis based on physical, macroeconomic, and financial fundamentals of oil prices. Section 4 concludes.

2. Background

The behaviour of the WTI crude oil futures price from June 13, 2006 to February 23, 2010 is pictured in Figure 1.

We observe that crude oil futures prices skyrocketed from 92$ a barrel in

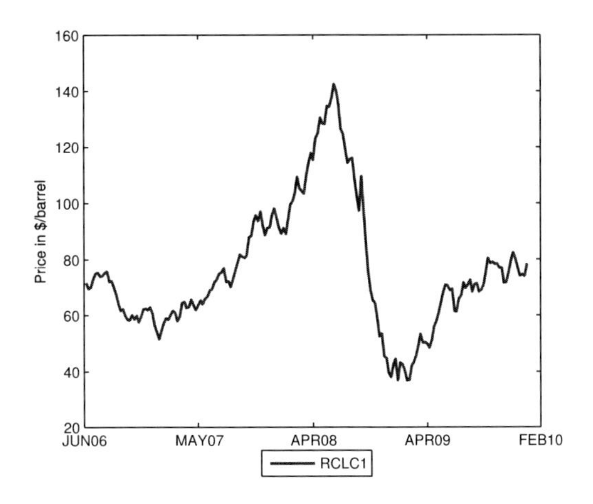

Figure 1. WTI Cushing, OK Crude Oil Future Contract 1 (Dollars per Barrel) from June 13, 2006 to February 23, 2010
Source: U.S. Energy Information Administration

January 2008 to cross the 140$ a barrel mark in June, finally hitting a record high of 147$ a barrel on July 11, 2008, before collapsing to less than 40$ a barrel in December.

Crude oil futures markets have been transforming radically over the last ten years following the U.S. Commodity Future Modernization Act (CFMA). The CFMA was approved by the U.S. Congress on December 15, 2000 and signed by President Clinton on December 21, 2000. Compared to previous regulation, it introduced more flexibility, allowing new financial agents (such as commodity index funds and market agents swapping commodities) to enter crude oil futures markets. The CFMA has notably withdrawn swap transactions from position limits previously established by the CFTC, and has fostered the emergence of contracts in differences (see Medlock and Jaffe (2009) for more details).

The years between 2006 and 2008 were marked by an explosion in the demand for oil sustained by strong world economic growth, both in emerging countries and in the U.S. Prices soared towards 100$ without affecting world economic growth too severely. At the same time, there was an upsurge in financial markets for oil, refined products and, more generally, for commodities. This rapid growth of the financial sphere goes hand in hand with increasing numbers of participants, financial products and marketplaces, some regulated (organised markets) and others, of increasing importance, unregulated OTC markets.

The period 2008-2009 has therefore raised the problem of interactions between the physical and financial fundamentals. It is marked by three successive phases: between January and July 2008, oil prices rose to 145$, which quickly raised questions over the potential role played by financial markets; between July and December 2008, they dropped to 36$, due to a financial adjustment in investors' positions and falling demand resulting from the economic crisis; during 2009, prices rose to 80$ which seems contrary to the state of the physical fundamentals, notwithstanding OPEC's production cuts.

In the next section, we attempt to establish clearly the links between crude oil futures prices and their fundamentals based on the best available statistical data.

3. Econometric Analysis

The econometric analysis contains first a brief review of studies (mostly in the professional literature) dealing with CFTC data, and then the regression results with physical, macroeconomic, and financial fundamentals.

3.1. Recent Econometric Studies using CFTC Data

The CFTC has been publishing more than three years of disaggregated data history since 4 September 2009 included in the weekly CoT report. For the oil market, these data focus on the NYMEX Light Sweet Crude Oil futures contract and supplement data previously available on the CFTC website.

The advantage of being able to access disaggregated CFTC data is the distinction between 'commercial' and 'non-commercial' traders. The boundary between these two types of trader is extremely blurred, however. There are many uncertainties over the classification made by the CFTC based on the raw data it has available. The CFTC establishes each trader as being 'mainly' commercial or non-commercial against its own criteria. These classifications based on raw data escape the public sphere, however, denying the academic community the opportunity to review these classifications, for example.

Current statistical studies covering the public CFTC data are restricted by the quality of the data themselves. The following, deliberately non-exhaustive, list includes studies published in 2008-2009.

3.1.1. Saporta et al. (2009)

The Bank of England report prepared by Saporta et al. (2009) is a full and recent statistical analysis of potential causes of the March-August 2008 shock in the oil market. The authors use public CFTC data on the long position of non-financial traders for their own statistical analysis (based on vector autoregressive models) of the 2003-2006 and 2006-2008 oil price samples. They conclude very cautiously that the hypothesis of speculative bubble (generated by non-commercial traders) cannot be ruled out totally when explaining the oil price dynamics of 2008 and highlight the inherent limitations in using public CFTC data. Therefore, the authors underline very clearly the advantage in accessing the CFTC[4] confidential data on investor positions, thereby identifying 'speculator' flows more accurately. Büyüksahin et al. (2008) have accessed these data for the study discussed previously.

3.1.2. Till (2009)

Till (2009) focuses on the issue of 'excessive speculation' in the oil market by also calculating the Working (1960) T index and by following the analysis by Sanders et al. (2010). The author concludes that for the heating oil and gasoline (petrol) NYMEX futures, the calculated Working indices fall within the average of other agricultural futures markets, where speculation is not rated as 'excessive'. Between summer 2007 and summer 2008, including options prices, Till indicates that the WTI NYMEX futures market became more speculative, despite the data indicating that it had reached an 'excessive' threshold as defined by Working (1960) and Sanders et al. (2010). By removing options data, futures taken in isolation could indicate excessive speculation in the oil market in the United States. Till remains very cautious, however, as to the conclusions of this

[4]The authors also mention a memo from the British Cabinet Office (2008), suggesting that '*without the major financial flows in the oil futures markets, the prices may not have risen then fallen so radically in 2008*'.

study: it should, for example, be widened to futures spreads to extend further the analysis of excessive speculation using the Working (1960) T index.

Having reviewed the existing studies, we develop in the next section our own econometric analysis of the 2008 oil price swing by introducing the physical, macroeconomic and financial fundamentals of oil prices gathered in our database. Then, we present the estimation results.

3.2. Part I: Physical Fundamentals

Let us start with the *physical* fundamentals of crude oil prices. Detailed information on the data used in the chapter, as well as their source, may be found in Table 1. Descriptive statistics are given in Table 2 during the study period going from June 13, 2006 to February 23, 2010.

On the supply side, spare capacity serves as an important indicator of market tightness because it shows how much supply can theoretically increase within a short time horizon, faced with growing demand or a supply disruption. According to data from the U.S. Energy Information Administration (see Table 1), the Organization of Petroleum Exporting Countries (OPEC) effective spare capacity was stable at around 2.5 million barrels per day in April 2008, but during the following months it fell sharply to 1.5 million barrels per day in July[5]. With low spare capacity, market participants can no longer rely on increased production from OPEC countries to fully offset disruptions and restore balance to the market without the need for significant price changes. The reduced level of spare production capacity therefore significantly increases the risk to oil prices from a disruption to supply because as many as 20 different countries currently produce at least 1 million barrels per day, including Iran, Iraq, Nigeria and Venezuela.

In parallel, increases in non-OPEC oil production capacity have been struggling over the period to keep pace with rapidly growing demand, particularly in China, the other emerging economies in Asia, and the U.S. Namely, we identify in the U.S. EIA data a declining trend in oil production in some key non-OPEC countries such as the U.S., the U.K. and Mexico while oil demand continued to grow in response to continued worldwide economic growth during the period.

[5] Since this period, OPEC spare capacity has continuously risen: it has more than quadrupled by April 2009, largely due to the global economic downturn and steadily weaker demand.

Table 1. List of Variables

Name	Variable	Source
RCLC1	WTI Cushing, OK Crude Oil Future Contract 1 (Dollars per Barrel)	U.S. Energy Information Administration
PRODMERC	Producer/Merchant/Processor/User Net Positions (Contracts of 1,000 Barrels)	U.S. Commodity Futures Trading Commission
SWAP	Swap Dealer Net Positions (Contracts of 1,000 Barrels)	U.S. Commodity Futures Trading Commission
SWAPTS	Swap Dealer Time Spread (Contracts of 1,000 Barrels)	U.S. Commodity Futures Trading Commission
MONEYM	Money Manager Net Positions (Contracts of 1,000 Barrels)	U.S. Commodity Futures Trading Commission
MONEYMTS	Money Manager Time Spread (Contracts of 1,000 Barrels)	U.S. Commodity Futures Trading Commission
OTHERREPT	Other Reportables Net Positions (Contracts of 1,000 Barrels)	U.S. Commodity Futures Trading Commission
OTHERREPTTS	Other Reportables Time Spread (Contracts of 1,000 Barrels)	U.S. Commodity Futures Trading Commission
WORKINGT	Working T Index	Authors
WRPUPUS2	Weekly U.S. Petroleum Products Product Supplied (Thousand Barrels per Day)	U.S. Energy Information Administration
WCESTUS1	Weekly U.S. Crude Oil Ending Stocks Excluding SPR (Thousand Barrels)	U.S. Energy Information Administration
WTTNTUS2	Weekly U.S. Total Crude Oil and Petroleum Products Net Imports (Thousand Barrels per Day)	U.S. Energy Information Administration
WCRFPUS2	Weekly U.S. Crude Oil Field Production (Thousand Barrels per Day)	U.S. Energy Information Administration
WCRRIUS2	Weekly U.S. Crude Oil Inputs into Refineries (Thousand Barrels per Day)	U.S. Energy Information Administration
WGFRPUS2	Weekly U.S. Finished Motor Gasoline Production (Thousand Barrels per Day)	U.S. Energy Information Administration
OPECSPARECAP	OPEC Crude Oil Spare Capacity (Million barrels per day)	U.S. Energy Information Administration
OPECSUPPLY	OPEC Crude Oil and Liquid Fuels Supply (Million barrels per day)	U.S. Energy Information Administration
NONOPEC SUPPLY	Non OPEC Countries Crude Oil and Liquid Fuels Supply (Million barrels per day)	U.S. Energy Information Administration
WORLDSUPPLY	Total World Production of Crude Oil (Million barrels per day)	U.S. Energy Information Administration
OECDCONS	OECD Countries Consumption of Crude Oil (Million barrels per day)	U.S. Energy Information Administration
NONOECDCONS	Non OECD Countries Consumption of Crude Oil (Million barrels per day)	U.S. Energy Information Administration
CHINACONS	China Consumption of Crude Oil (Million barrels per day)	U.S. Energy Information Administration
OTHERASIACONS	Other Asian Countries Consumption of Crude Oil (Million barrels per day)	U.S. Energy Information Administration
WORLDCONSU	Total World Consumption of Crude Oil (Million barrels per day)	U.S. Energy Information Administration
USCOMINV	US Commercial Inventory (Million barrels)	U.S. Energy Information Administration
OECDCOMINV	OECD Commercial Inventory (Million barrels)	U.S. Energy Information Administration
SP500	S&P 500 INDEX	Thomson Financial Datastream
VIX	CBOE Volatility Index VIX	Chicago Board Options Exchange
GSENSPT	S&P GSCI Energy Spot- PRICE INDEX	Thomson Financial Datastream
DJUBHOT	DJ UBS-Heating Oil Sub Index TR - RETURN IND. (OFCL)	Thomson Financial Datastream
EXRUSDEUR	ECB reference exchange rate, US dollar/Euro, 2:15 pm (C.E.T.)	European Central Bank
CPIENGNS	Consumer Price Index for All Urban Consumers: Energy	U.S. Department of Labor: Bureau of Labor Statistics
PPIFEG	Producer Price Index: Finished Energy Goods	U.S. Department of Labor: Bureau of Labor Statistics
PPIACO	Producer Price Index: All Commodities	U.S. Department of Labor: Bureau of Labor Statistics
TB1YR	1-Year Treasury Bill: Secondary Market Rate	Board of Governors of the Federal Reserve System
IPMAT	Industrial Production: Materials	Board of Governors of the Federal Reserve System
UMCSENT	University of Michigan: Consumer Sentiment	Survey Research Center: University of Michigan
MICH	University of Michigan Inflation Expectation	Survey Research Center: University of Michigan
GDP	Gross Domestic Product, 1 Decimal	U.S. Department of Commerce: Bureau of Economic Analysis
BUSINV	Inventories: Total Business	U.S. Department of Commerce: Census Bureau
CBI	Change in Private Inventories	U.S. Department of Commerce: Bureau of Economic Analysis
PAYEMS	Total Nonfarm Payrolls: All Employees	U.S. Department of Labor: Bureau of Labor Statistics
DUMMYGSMS	Dummy Goldman Sachs Morgan Stanley News Announcements	Authors
DUMMYREFIN	Dummy Crisis Refining	Authors
DUMMYSAUDIARABIA	Dummy Saudi Arabia News Announcements	Authors

, Germany, Greece, Hungary, Iceland, Ireland, Italy, Japan, Luxembourg, Mexico, the Netherlands, New Zealand, Norway, Poland, Portugal,Slovakia, South Korea, Spain, Sweden, Switzerland, Turkey, the United Kingdom, and the United States.
$OPEC$ = Organization of Petroleum Exporting Countries: Algeria, Angola, Ecuador, Iran, Iraq, Kuwait, Libya, Nigeria, Qatar, Saudi Arabia, the United Arab Emirates, Venezuela.

With futures markets in contango, market participants have a strong demand for inventories, as there emerges clear opportunities for arbitrageurs to buy spot and sell forward[6]. Data from the U.S. EIA (as detailed Table 1) reveals some inventory builds (roughly 0.6 million barrels per day) during April-June 2008 in non-OECD regions, in addition to moderate stockbuilds in China preceeding the Olympic Games. Global crude oil inventories started to accumulate in the second quarter 2008, with the pace of build picking up after August as demand and prices began to slide[7].

Finally, worldwide refining sectors bottlenecks have implications for crude oil futures prices. Excess capacity in the refining industry has been shrinking as demand has grown, and has left less of a buffer for periods when supply and demand becomes tight (as during the period considered). Besides, a tight availability of light sweet crude oil, most notably from Nigeria, and recently-introduced mandates on low sulphur fuel standards in OECD economies leading to greater interest by refiners to run light sweet crude oil during 2007 and the first half of 2008 are a clear case of a tightening supply and demand balance for that segment on the crude oil market.

Knowing the very low price responsiveness of both the demand and supply of oil[8], the interpretation according to the *physical* fundamentals defenders unfolds as follows. As world GDP growth during 2007 and through the first half of 2008 was very strong by historical standards, and supply was more or less fixed by capacity constraints[9], oil prices would be expected to rise substantially. When the global economy started to slow down and then entered into recession

[6]Cash-and-carry or storage arbitrage favour stockbuilds, and therefore any price above or below the equilibrium warranted by supply and demand must be explained by an act of hoarding in the form of inventory builds or withholding production.

[7]Note that in the case of oil, the data on oil inventories are notoriously poor, with many countries not reporting at all. In particular, most non-OECD countries, which make up a little less than half of world demand for crude oil and include very large consumers such as China, do not report data on oil inventories. Furthermore, oil inventories do not include oil in tankers, commonly referred to as 'oil at sea', which distorts even the inventories data reported by the United States and other OECD countries.

[8]The short-run price elasticity of demand is estimated to be less than -0.1 and the long-run price elasticity ranges between -0.2 and -0.3 (Hamilton (2009)).

[9]Total OPEC production rose steadily month by month from January through July 2008, and average daily production during these seven months was about 4 percent above the average daily production in 2007.

Table 2. Summary Statistics from June 13, 2006 to February 23, 2010

	Mean	Median	Maximum	Minimum	Std. Dev.	Skewness	Kurtosis	Jarque-Bera	Observations
BUSINV	-1.83E-05	0.000497	0.002461	-0.003614	0.001572	-0.797617	2.422166	23.14926	193
CBI	-0.553754	0.002429	18.49812	-19.66844	7.960997	0.000320	2.934262	0.034755	193
CHINACONSUMPTION	0.000481	0.000396	0.038420	-0.032185	0.013508	0.156335	3.208458	1.135621	193
CPIENGNS	-0.000156	0.001517	0.022252	-0.041563	0.011685	-1.296619	5.367294	99.14539	193
DJUBHOT	-0.002061	-0.000418	0.203610	-0.144108	0.051426	0.296050	4.265238	15.69259	193
EXRUSDEUR	0.000418	0.001261	0.100764	-0.067992	0.017059	0.447509	10.36070	442.1383	193
GDP	0.002474	0.002491	0.005325	-0.000293	0.001097	0.021172	3.105474	0.103881	193
GSENSPT	0.000180	0.004306	0.193277	-0.140660	0.049933	-0.027122	4.031309	8.576757	193
IPMAT	-0.000599	-0.000262	0.007570	-0.014412	0.003048	-1.762970	8.764100	367.1592	193
MICH	-0.001664	-0.000878	0.080507	-0.126858	0.030994	-0.875657	6.624692	130.3192	193
MONEYM	636.3005	964.0000	38639.00	-43849.00	15894.32	-0.108715	2.599871	1.667670	193
MONEYMTS	143.2902	-268.0000	120363.0	-30597.00	13033.30	3.932915	39.04163	10943.67	193
NONOECDCONS	0.000489	0.000240	0.014013	-0.009369	0.003887	0.540557	3.928802	16.33650	193
NONOPECSUPPLY	9.00E-05	4.58E-05	0.008109	-0.008130	0.002210	-0.008173	6.361191	90.85372	193
OECDCOMINV	0.000251	0.000462	0.005524	-0.005664	0.002313	-0.439895	2.929755	6.264171	193
OECDCONS	-0.000513	-0.000607	0.009847	-0.010691	0.004517	0.021543	2.354114	3.369661	193
OPECSPARECAP	0.006095	0.002181	0.080717	-0.047966	0.026440	0.604173	3.519055	13.90822	193
OPECSUPPLY	-0.000122	-1.41E-05	0.004865	-0.006910	0.002097	-0.517983	3.965620	16.12876	193
OTHERASIACONSUMP	0.000216	0.000511	0.010029	-0.009640	0.004298	-0.334754	2.672045	4.469527	193
OTHERREPT	290.9378	-1683.000	127973.0	-70846.00	16484.64	2.160745	21.81048	2995.598	193
OTHERREPTTS	268.4611	1308.000	41139.00	-55330.00	13068.92	-0.683544	4.787219	40.71562	193
PAYEMS	-0.000225	-9.18E-05	0.000513	-0.001303	0.000487	-0.633533	2.271468	17.17872	193
PPIACO	0.000246	0.001163	0.006974	-0.013261	0.004065	-1.154935	4.355982	57.69245	193
PPIFEG	-2.79E-05	0.001882	0.020577	-0.034655	0.010379	-1.066723	4.377650	51.86480	193
PRODMERC	-718.3005	-1090.000	34105.00	-57323.00	10961.93	-0.253255	6.522390	101.8379	193
RCLC1	0.000465	0.005684	0.155272	-0.187231	0.050483	-0.488192	4.436080	24.25086	193
SP500	0.000694	-0.000633	0.200837	-0.113559	0.033429	0.950413	10.12471	437.2630	193
SWAP	364.8601	-609.0000	48094.00	-45947.00	14351.21	0.127496	3.559961	3.044388	193
SWAPTS	246.1865	737.0000	21830.00	-38702.00	10985.44	-0.580950	3.622201	13.96956	193
TB1YR	-0.013004	-0.005689	0.083788	-0.186508	0.035982	-1.856089	10.78341	597.9917	193
UMCSENT	-0.001071	-0.001686	0.031194	-0.050266	0.014381	-0.339202	3.550889	6.141503	193
USCOMINV	0.000352	0.001326	0.006186	-0.013071	0.003545	-1.139872	4.595075	62.25454	193
VIX	0.000635	-0.009569	0.564734	-0.369765	0.134197	0.584069	4.897380	39.92366	193
WCESTUS1	-0.000123	0.000282	0.031173	-0.028113	0.011416	-0.020016	2.801215	0.330657	193
WCRFPUS2	0.000284	0.000586	0.141380	-0.221575	0.023963	-3.006829	45.57322	14866.17	193
WCRRIUS2	-0.000597	-0.000810	0.118889	-0.140321	0.023400	-0.920106	15.62379	1308.752	193
WGFRPUS2	-0.000198	0.001841	0.088498	-0.117598	0.024064	-0.901943	7.962239	224.1841	193
WORKINGT	-0.000501	-0.001564	0.057567	-0.085275	0.019934	-0.318299	4.506420	21.50791	193
WORLDCONS	-6.66E-05	0.000264	0.008042	-0.005938	0.002958	0.088138	2.374686	3.394310	193
WORLDSUPPLY	2.58E-05	-1.49E-05	0.004667	-0.005828	0.001698	-0.328077	4.153529	14.16274	193
WRPUPUS2	-0.000368	9.60E-05	0.064579	-0.062189	0.021718	-0.013464	3.501718	2.030086	193
WTTNTUS2	-0.001371	-0.002060	0.211304	-0.161887	0.066288	0.158320	3.144268	0.973640	193

in the second half of 2008, demand fell off rapidly, and since supply declined more slowly as the OPEC began its production cuts in mid-2008, prices dropped like a stone.

With such tightness in supply and demand fundamentals of crude oil, all prerequisites seems to be fulfilled to conduct to the 2008 'crisis' from the physical viewpoint of the market.

To test statistically this hypothesis, we will estimate in Section 3.5 a regression model with the following dependent variable:

- $RCLC1_t$ is the WTI Cushing, OK Crude Oil Future Contract at time t.

Besides, we will introduce the following independent variables as representative of 'physical' determinants in our regression model:

- $WRPUPUS2_t$ is the Weekly U.S. Petroleum Products Product Supplied,
- $WCESTUS1_t$ is the Weekly U.S. Crude Oil Ending Stocks Excluding Strategic Petroleum Reserves,
- $WTTNTUS2_t$ is the Weekly U.S. Total Crude Oil and Petroleum Products Net Imports,
- $WCRFPUS2_t$ is the Weekly U.S. Crude Oil Field Production,
- $WCRRIUS2_t$ is the Weekly U.S. Crude Oil Inputs into Refineries,
- $WGFRPUS2_t$ is the Weekly U.S. Finished Motor Gasoline Production,
- $OPECSPARECAP_t$ is the OPEC Crude Oil Spare Capacity,
- $OPECSUPPLY_t$ is the OPEC Crude Oil and Liquid Fuels Supply,
- $NONOPECSUPPLY_t$ is the Non OPEC Countries Crude Oil and Liquid Fuels Supply,
- $WORLDSUPPLY_t$ is the Total World Production of Crude Oil,
- $OECDCONS_t$ is the OECD Countries Consumption of Crude Oil,
- $NONOECDCONS_t$ is the Non OECD Countries Consumption of Crude Oil,

- $CHINACONS_t$ is the China Consumption of Crude Oil,
- $OTHERASIACONS_t$ is the Other Asian Countries Consumption of Crude Oil,
- $WORLDCONS_t$ is the Total World Consumption of Crude Oil,
- $USCOMINV_t$ is the US Commercial Inventory,
- $OECDCOMINV_t$ is the OECD Commercial Inventory,
- $DUMMY\ SAUDIARABIA$ is a dummy variable which takes the value of one during June 2008 when Saudi Arabia attempted to stabilize crude oil prices[10] and zero otherwise,
- $DUMMYREFIN$ is a dummy variable which takes the value of one during July-August 2008 when shortages in refining capacities were documented[11] and zero otherwise.

In the next section, we extend our empirical investigation to the *macroeconomic* determinants of crude oil futures prices.

3.3. Part II: Macroeconomic Fundamentals

Now, let us focus on the links between crude oil futures and *macroeconomic* determinants. From the macroeconomic viewpoint, the global recession played an important role in the subsequent collapse of crude oil prices.

With sustained low interest rates, the monetary policy of the U.S. Federal Reserve has contributed to pushing the U.S. dollar to 15-month low against a basket of major currencies.

As WTI crude oil futures prices are denominated in U.S. dollar, the exchange rate for the dollar changes oil prices in other currencies. Hence, a weaker dollar means lower prices in Euro, which should translate into higher oil demand in the Eurozone and tighten the global supply and demand balance, all other things being held equal.

[10]See for instance BBC NEws announcements 'Oil price up despite Saudi pledge' on June 23, 2008 at *http://newsvote.bbc.co.uk/*

[11]See for instance the ArabianBusiness news column "Refinery 'shortage' behind high oil price' on June 17, 2008 at *http://www.arabianbusiness.com/*

Another macroeconomic effect is linked to inflation. Investors shift funds into commodity markets as a hedge against inflation. Inflation in the U.S. would imply a weaker dollar, so inflationary trends or fears of inflation encourage buying of futures by financial investors.

After the recent economic turmoil, crude oil futures prices may be tracking upward moves in other markets (such as equities) and perceptions on economic recovery. That is why we use global indices of business sentiment as potential macroeconomic drivers in the subsequent analysis.

On top of physical fundamentals, the following independent variables are selected as being representative of macroeconomic factors in our regression model:

- $SP500_t$ is the S&P 500 INDEX,
- VIX_t is the CBOE Volatility Index VIX,
- $EXRUSDEUR_t$ is the US dollar/Euro ECB reference exchange rate,
- $CPIENGNS_t$ is the 'Energy' Consumer Price Index for All Urban Consumers,
- $PPIFEG_t$ is the "Finished Energy Goods" Producer Price Index,
- $PPIACO_t$ is the 'All Commodities' Producer Price Index,
- $TB1YR_t$ is the Secondary Market Rate 1-Year U.S. Treasury Bill,
- $IPMAT_t$ is the 'Materials' Industrial Production,
- $UMCSENT_t$ is the University of Michigan Consumer Sentiment,
- $MICH_t$ is the University of Michigan Inflation Expectation,
- GDP_t is the U.S. Gross Domestic Product,
- $BUSINV_t$ is the Total Business Inventories,
- CBI_t is the Change in Private Inventories,
- $PAYEMS_t$ is the 'All Employees' Total Nonfarm Payrolls.

In the next section, we finalize our empirical investigation by examining the potential influence of purely *financial* fundamentals in crude oil prices.

3.4. Part III: Financial Fundamentals

Financial players are often presented by market analysts as outweighing commercial or physical players, buying or selling far more paper barrels than physical players do. Although some financial institutions engage in physical trading of oil by controlling storage facilities, most financial players enter the oil market through futures contracts which do not result in physical delivery.

Commodities, oil in particular, have become a well-recognised asset class within investment portfolios of financial institutions as a means to diversify inherent risks of inflation, dollar depreciation (for U.S. investors) or equity market weakness. Investments into commodity indices have grown rapidly over the period: not only do they increase in value when commodity prices rise, and vice versa, they also reflect the roll return, that is the return from selling the futures contracts in the basket as they approach the expiration month and buying the same contracts in a deferred month. Each index has different rules for when and how the contracts are rolled forward. We use in this chapter the two most actively traded commodity indices, *i.e.* the S&P Goldman Sachs Commodity Index (GSCI) Energy Spot Price Index and the Dow Jones UBS Heating Oil Sub Index.

Speculation appears as another possible explanation for the 2008 crude oil price swing. While physical and macroeconomic fundamentals were responsible for the rise in the equilibrium oil price, speculation on the futures price of crude oil might have led to both overshooting of prices in the first half of 2008 and undershooting in the second half of the year. Unfortunately, it is very difficult to measure speculation in any direct way. One obvious straightforward approach is to look at the data on the open interest of oil futures, which have grown spectacularly for large investors over the last few years.

By 2008, daily trading in paper barrels had reached 15 times the daily world production of oil (of around 85 million barrels per day) and remained at about that level through the first half of 2009[12]. These numbers provide evidence of the enormous "financialization" of the oil market that has taken place since 2000, and some part of this may reflect speculative activities. As a reaction to this situation, the CFTC has announced on July 7, 2009 that it would con-

[12]Note that the futures data are probably an underestimate, since they do not include options or over-the-counter trades. Furthermore, it is important to note that major oil producers, like Saudi Arabia and the other OPEC countries, deal only in the spot market and not in the futures market

sider imposing position limits[13] on futures contracts for energy products, which would bring oil in line with the policy of limits the CFTC places on speculative trading in agricultural products like wheat and corn.

One main problem, however, arises when using indicators of speculative activity for crude oil futures. First, to isolate the speculative component in the net open positions, one has to separate out commercial and non-commercial positions, since the former are presumably related to hedging rather than speculation. While separate data on commercial and noncommercial net open positions are reported by the CFTC, in practice the distinction between the two is more or less arbitrary. The CFTC has pledged to improve the quality of futures data, as it released on September 4, 2009 sub-categories for the Commitment of Traders weekly report. According to the CFTC, the disaggregated CoT report increases transparency by separating traders into four categories. A 'Producer/Merchant/Processor/User' is an entity that predominantly engages in the production, processing, packing or handling of a physical commodity and uses the futures markets to manage or hedge risks associated with those activities. A 'Swap Dealer' is an entity that deals primarily in swaps for a commodity and uses the futures markets to manage or hedge the risk associated with those swaps transactions. The swap dealer's counterparties may be speculative traders, like hedge funds, or traditional commercial clients that are managing risks arising from their transactions in the physical commodity. A 'Money Manager' is a registered Commodity Trading Advisor (CTA), a registered Commodity Pool Operator (CPO), or an unregistered fund identified by the CFTC. These traders are engaged in managing and conducting organized futures trading on behalf of clients. Finally, the 'Other Reportable' category concerns traders that are not placed into one of the three other categories.

More than three years of historical data for the new Disaggregated Commitment of Traders report were subsequently released by the CFTC. While the CFTC aims at classifying participants according to the nature of their trading, practical constraints prevent it from capturing in depth the precise essence of each entity's dealings, especially those of swap dealers, who 'swap' derivatives contracts tailored to customer needs and hedge the associated risks on the fu-

[13] So far, futures exchanges themselves, principally NYMEX, have been allowed to set their own limits.

tures exchange[14].

Our analysis of the financial fundamentals in crude oil futures prices is mainly based on CFTC disaggregated data, which yields to considering the following independent variables:

- $GSENSPT_t$ the S&P GSCI Energy Spot Price Index,
- $DJUBHOT_t$ the DJ UBS-Heating Oil Sub Index,
- $PRODMERC_t$ the Net Positions of the CFTC 'Producer/Merchant/ Processor/User' Category,
- $SWAP_t$ the Net Positions of the CFTC 'Swap Dealer' Category,
- $SWAPTS_t$ the 'Swap Dealer' Time Spread,
- $MONEYM_t$ the Net Positions of the CFTC 'Money Manager' Category,
- $MONEYMTS_t$ the 'Money Manager' Time Spread,
- $OTHERREPT_t$ the Net Positions of the CFTC 'Other Reportables' Category,
- $OTHERREPTTS_t$ the 'Other Reportables' Time Spread,
- $WORKINGT_t$ the Working T Speculative Index,
- $DUMMYGSMS$ a dummy variable which takes the value of one during crude oil news investments by Goldman Sachs and Morgan Stanley[15] and zero otherwise.

The Working T Speculative Index is computed as $T = 1 + SS/HS$ with SS the Speculators (*i.e.*, the CFTC CoT 'Money Managers' categories) Short Positions and HS the Hedgers (*i.e.*, the CFTC CoT 'Producers' and 'Swap Dealers' categories) Short Positions. To deal with the 'Other Reportables' portion of open interest, we have allocated their positions to 'Speculators' and 'Hedgers' according to their respective share of the overall reported positions. In the spirit of Working (1960), futures market are indeed used primarily for hedging, and

[14]Notably, a substantial portion of passive investors are known to gain desired exposures to commodities markets through swap dealers.

[15]See the Appendix for more details.

speculation can only be judged as 'excessive' relative to the level of hedging activity in the market.

We conduct in the next section the estimation of the econometric model with regards to the physical, macroeconomic and financial fundamentals of crude oil futures prices.

3.5. Estimation Results

3.5.1. Econometric strategy

In this section, we provide more details on the implementation of the econometric procedures and tests. From the database of physical, macroeconomic and financial fundamentals detailed in Sections 3.2 to 3.4, the econometric strategy adopted is referred to as 'backward elimination':

1. we estimate the full model with all independent variables descrived above by OLS with Newey-West Heteroskedasticity and Autocorrelation Consistent (HAC) covariance matrix estimators,
2. we withdraw one by one non significant variables (by starting with the least significant variable) and re-estimate the model,
3. we obtain a restricted model with only statistically significant independent variables at usual confidence levels.

Let us discuss here the preliminary concern with possible colinearity between independent variables. Indeed, if the independent variables in the model are highly correlated (multicollinearity), the reported regression coefficients may be severely distorted and thus the results are not reliable. First, we have investigated the matrix of partial cross-correlations between independent variables[16]. We have verified that the correlation coefficients are not higher than 0.6 as usually retained as the threshold for colinearity. Second, since it is possible to have low correlations together with colinearity, we have investigated the presence of multicolinearity by computing the inflation of variance between independent variables. These calculations did not reveal either serious problematic multicolinearities[17].

---

[16] This matrix is not reported in the chapter to conserve space, and may be obtained upon request.

[17] Similarly, these results are not reported but may be obtained upon request.

Finally, this 'backward elimination' procedure allows us to identify the following restricted model:

$$\begin{aligned} RCLC1_t = {} & \alpha + ar(1) + ma(1) + \beta_1 OPECSPARECAP_t \\ & + \beta_4 OTHERASIACONS_t + \beta_5 GSENSPT_t \\ & + \beta_6 PPIFEG_t + \beta_7 GDP_t + \beta_8 UMCSENT_t \\ & + \beta_9 MONEYM_t + \beta_{10} PRODMERC_t + \beta_{11} SWAP_t \\ & + \beta_{12} DUMMYREFIN + \beta_{13} WORKINGT_t + \epsilon_t \end{aligned} \quad (1)$$

with α the constant term, $ar(1)$ and $ma(1)$ autoregressive and moving average terms of order 1, ϵ_t the error term, and all other variables that were defined previously.

All variables have been transformed to stationary before estimating eq(1)[18]. In order to match the weekly frequency of the CFTC data, all variables are presented in weekly frequency[19] from June 13, 2006 to February 23, 2010. Results of eq(1) are shown in Table 3.

As explained above, only reduced-form estimates are presented here, *i.e.* non-significant variables were dropped one by one from the complete model detailed in eq(1), and the reduced-form model was re-estimated until all variables are statistically significant. All diagnostic tests are valid for the estimation of eq(1), *i.e.* the Ljung-Box-Pierce test indicates that the residuals are not autocorrelated, and, as judged by the F-statistic p-value, the joint significance of results is accepted at the 1% significance level. Following the Box-Jenkins methodology, we configure the data generating process for the oil variable as being an ARMA(p,q) of order 1. Indeed, we observe in Table 3 that the autoregressive ($AR(1)$) and moving average ($MA(1)$) coefficients of order 1 are statistically significant.

3.5.2. Estimates for physical fundamentals

Among the *physical* fundamentals of oil prices, we identify two independent variables as driving crude oil futures prices. On the supply side,

[18] Standard unit root test results (ADF, PP, KPSS) are available upon request to the authors.

[19] Note variables which were initially in monthly frequency have been linearly interpolated by using the Matlab function by L.Shure which minimizes the mean square error between the original data and their ideal values.

Table 3. Regression Results for WTI Crude Oil Price *Physical, Macroeconomic and Financial* Determinants

Dependent Variable: RCLC1

Variable	Coefficient	Std. Error	t-Statistic	Prob.
Constant	0.001666	0.000922	1.807157	0.0724
AR(1)	0.459701	0.034780	13.21729	0.0000
MA(1)	0.988354	0.005790	-170.6927	0.0000
OPECSPARECAP	0.065560	0.013143	4.988144	0.0000
GSENSPT	0.589640	0.042020	14.03225	0.0000
GDP	-0.943992	0.393878	-2.396663	0.0176
UMCSENT	-0.091031	0.042527	-2.140557	0.0336
MONEYM	2.44E-07	1.48E-07	1.655578	0.0995
WORKINGT	0.236644	0.124331	1.903334	0.0586
DUMMYREFIN	0.002789	0.001632	1.708553	0.0892

Diagnostic Tests	
Adjusted R-squared	0.615689
Ljung-Box-Pierce Test	0.191
Akaike info criterion	-4.036616
Schwarz criterion	-3.866955
Durbin-Watson statistic	2.060907
F-statistic (p-value)	0.000000

$OPECSPARECAP_t$ is significant and positive at the 1% level. Thus, we are able to identify that the OPEC Crude Oil Spare Capacity as a key variable driving crude oil futures prices over the period. Its positive sign indicates that more supply of crude oil into the market was translated into price *increases* in a context of limited supply. This variable is pictured in Figure 2: it conveys the information that dramatic changes occurred in the *physical* fundamentals of the crude oil futures market around July-August 2008.

Besides, we uncover the statistically significant impact of $DUMMYREFIN$ at the 10% level. The sign is also positive, which suggests that refinery bottlenecks during July-August 2008 did contribute to crude oil futures price increases. This variable therefore brings us a better understanding of the geopolitical explanations behind crude oil price changes (namely a mismatch between

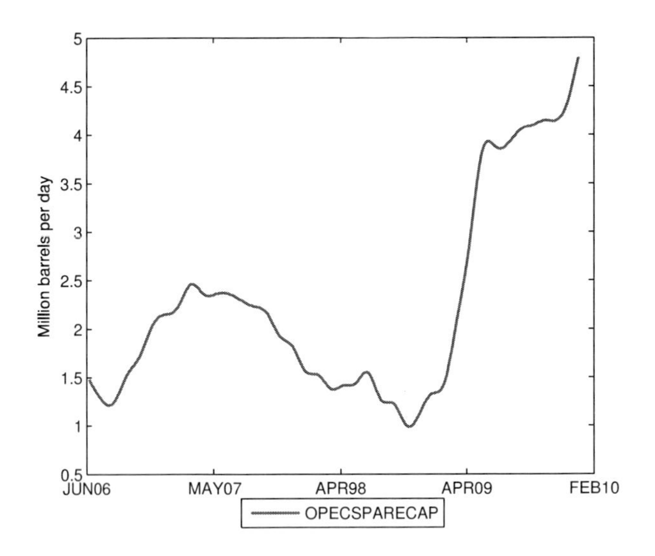

Figure 2. *Physical* Determinants of WTI Crude Oil Price from June 13, 2006 to February 23, 2010
Source: U.S. Energy Information Administration

crude availability and refinery capacity as commented above).

However, on the demand side, $OTHERASIACONS_t$ is not statistically significant. We were expecting that more consumption of crude oil in other Asian Countries would drive price increases, in a context of sustained economic growth until the recession mid-2008. Therefore, we are unable to verify statistically that the "tightness" of the market in terms of limited physical delivery *vs.* strong demand constitutes one of the main explanations behind crude oil price changes over the period, as advocated by many analysts.

3.5.3. Estimates for macroeconomic fundamentals

Among the *macroeconomic* determinants of crude oil futures prices, we identify in Table 3 the statistical influence of GDP_t and $UMCSENT_t$ at the 5% level. Therefore, the U.S. Gross Domestic Product and the University of Michigan Consumer Sentiment are identified as being the key macroeconomic determinants of crude oil futures prices over the period. Their *negative* sign suggests that the economic downturn has fostered price *decreases* on the crude oil futures market. Thus, given the perception of sustained low interest rates in the U.S., weak macroeconomic fundamentals are shown to influence statistically oil price changes.

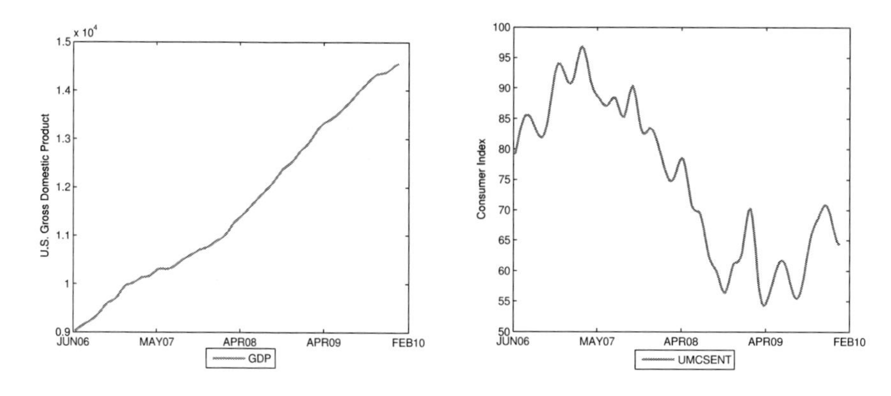

Figure 3. *Macroeconomic* Determinants of WTI Crude Oil Price from June 13, 2006 to February 23, 2010
Source: Thomson Financial Datastream, U.S. Department of Labor: Bureau of Labor Statistics, U.S. Department of Commerce: Bureau of Economic Analysis, Survey Research Center: University of Michigan

As shown in Figure 3, these variables show evidence that economic growth and global demand were progressively shrinking from August 2008 onwards. Only the GDP variable has been decreasing more gradually, until market analysts recognized that the economy entered a period of 'recession'.

3.5.4. Estimates for financial fundamentals

Among the *financial* determinants of crude oil futures prices, we uncover in Table 3 the statistical significance of $GSENSPT_t$ at the 1% level. Among CFTC disaggregated data, only one category of market participants is found to be statistically significant: $MONEYM_t$ at the 10% level. The *positive* signs of the GSCI Energy Spot Price Index, as well as the Net Positions of the CFTC 'Money Manager' category concur to the view that purely financial fundamentals may have fostered crude oil futures price *increases* over the period 2006-2008, ending up in the quasi-bubble situation of the summer 2008. In addition, the $WORKINGT_t$ index is significant in this regression at approximately the 5% level, which suggests the presence of 'excessive' speculation over the period.

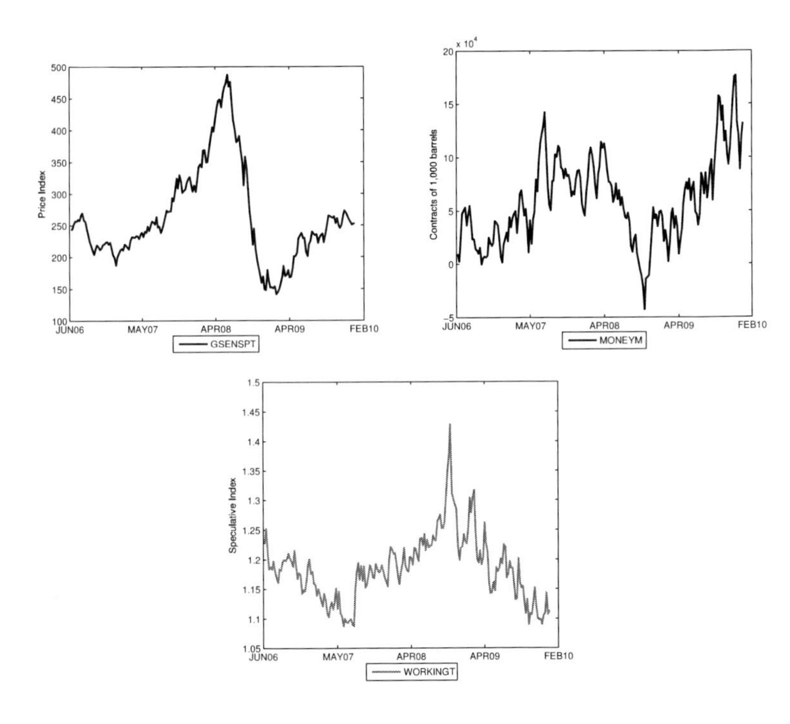

Figure 4. *Financial* Determinants of WTI Crude Oil Price from June 13, 2006 to February 23, 2010
Source: U.S. Commodity Futures Trading Commission

$GSENSPT_t$ is picured in Figure 4, along with the significant disaggregated CFTC variable and the Working T index. We observe visually that the commodity price index $GSENSPT_t$ exhibits dramatic changes before and after the summer 2008, thereby illustrating the connections between global commodity markets (including crude oil futures) and macroeconomic conditions.

Concerning the CFTC variable, we notice that $MONEYM_t$ varies contra-cyclically with respect to crude oil price changes. This *à priori* puzzling result shows that money managers enter the market to *anticipate* changes in fundamentals, thereby showing that the market works efficiently. Thus result appears in line with the view that index investors and fund managers may be able to anticipate future changes in the supply and demand conditions in the crude oil futures market.

Last but not least, the Working T Speculative Index clearly exhibits a peak during the summer 2008, which highlights the possibility of 'excessive' speculation during that period.

3.5.5. Summary of the estimation results

Overall, we confirm that several different factors seem to have caused the increase in crude oil futures prices during the period.

Among *physical* fundamentals, the OPEC Crude Oil Spare Capacity and the dummy variable for July-August 2008 bottlenecks in the refining industry are found to be the most significant drivers. Tight spare capacity and the lack of complex refining capacity are therefore confirmed as having a good explanatory power in crude oil futures prices *ceteris paribus*.

Among *macroeconomic* fundamentals, the University of Michigan Consumer Sentiment and the U.S. Gross Domestic Product reflect the significant influence of changing business conditions on crude oil futures prices.

Interestingly, among *financial* fundamentals, the S&P GSCI Energy Spot Price Index, the Net Positions of the CFTC 'Money Manager' Category and the Working T index are found to be significant in explaining crude oil futures price changes over the period.

Collectively, we obtain robust empirical results: financial factors remain significant in explaining crude oil futures prices changes, after controlling for other physical markets and macroeconomic conditions.

By aggregating physical, macroeconomic and financial determinants, we showed that price changes of the scale and the significance of the 2008 episode do not result from a single cause, but rather from a confluence of factors. The extraordinarily strong global economic growth until 2008, the constrained oil supply from key producers (Russia, Venezuela, Iran, Nigeria, Iraq, etc.), the lack of OPEC spare production capacity and the lack of spare refining capacity to handle heavy sour crude oil all played in role in driving crude oil futures prices to 147$ per barrel. With highly inelastic supply and demand, the influence of financial investors is found to contribute just as much to price changes as physical and macroeconomic determinants.

Speculation may therefore play one role among many in influencing prices. Increases in speculative activity in crude oil futures markets (as measured by open interest net positions of large investors) may be seen indeed as a result of the high level of oil prices and the high uncertainty surrounding the value of future oil prices, not the other way around. In times of ample spare capacity, there is little motivation for commercial producers and users of energy to hedge, since there is little perceived risk. As a consequence, there is only a small role for those who whish to take on the risk, *i.e.* speculators. In contrast, when excess capacity declined and market participants perceived that OPEC members would no longer maintain stable prices in the environment of geopolitical risks, market participants became increasingly less certain of the path of future oil prices. This increased uncertainty caused commercial producers and energy users to increase their desire to hedge. Hence, there has been a much larger role for those prepared to bear the risks in the markets, *i.e.* the speculators.

These results tend to confirm the findings by Sornette et al. (2009), who also find evidence of speculation during the 2006-2008 oil bubble in a different econometric framework based on econophysics. However, they do not rely on CFTC data to strenghten their results, as in this chapter.

Conclusion

This chapter seeks to make three contributions: *(i)* an analysis of the types of factors that are known to affect crude oil prices, *(ii)* the use of the CFTC data to detect speculative behavior, and *(iii)* estimate the model with physical, macroeconomic and financial fundamentals. Moreover, we are primarily interested in the recent, 2006-2010, price changes and volatility in the oil futures market (price run-up in 2008, or the potential that there was a bubble in 2008), and in determining if the state of the market has changed for oil prices (or at least for oil futures).

The growth in financial instruments since the U.S. derivatives markets reforms of the 2000s has certainly aided price discovery in various commodities markets (including crude oil futures) by bringing into the market extra information on current and future market conditions. The 2008 massive spike in oil prices in the midst of the deepest recession since 1945 has led to a suspicion that financial speculation has become the prime driver of prices. Hence, our analysis has been focusing on the possible emergence of new *financial* fundamentals in crude oil futures prices, without overlooking other economic determinants.

From our complete analysis *including physical markets and macroeconomic effects*, it appears reasonable to conclude that large exogenous changes in "Money managers" net positions have been partly driving higher crude oil futures prices over the period. On the one hand, speculation by financial actors has amplified the upwards or downwards price movements, as explained by changes in demand and supply fundamentals. On the other hand, speculators are responding both rationally and pre-emptively to economic conditions, creating considerable liquidity and offering information to the physical market in the process, which appears vital to provide the correct incentives for future investments.

A few caveats are necessary. One current flaw of the new CFTC disaggregated version of the Commitment of Traders report is that it does not fully capture information about the swap dealers' counterparties. Besides, we lack further dimensions (as in Büyükşahin et al. (2008)) from CFTC privately-owned data, which opens the contentious issue of defining market agents' categories from 'raw' data.

It cannot be excluded that such movements occur again in the years to come, with natural volatility joined by that of financial investors who consider oil (and more generally commodities) as a class of arbitrable assets compared with others. Strong pressure on the prices will appear by the end of the decade, mainly due to physical fundamentals (under-investment in new production capacities). The functioning of financial oil markets and the financial logic of their operators include risks which are difficult to control and may generate a systemic risk.

The question of the price of oil therefore ends in the more general problem of financial market regulation. The policies being considered by the CFTC to put aggregate position limits on futures contracts and to increase the transparency of futures markets are moves in the right direction. Among widespread calls for tighter controls on speculative trading in futures markets, the creation of clearinghouses for all transactions in derivatives trading would ensure that more captial would have to be set aside or increased margins would be required for trading derivatives contracts.

Appendix: Goldman Sachs and Morgan Stanley News Energy (including Crude Oil) Investments

Date	Annoucement	Source
13/06/06	MS spends 3 billion to buy 38	
01/12/06	Goldman Sachs Environmental Policy: 2006 Year-End Report	Goldman Sachs
01/01/07	Goldman Sachs negotiating to buy 30% of Solel Solar	IVC Weekly
01/01/07	Secures Project Financing of US$100 Million with Morgan Stanley for Blue Mountain's Faulkner I Power Plant	Nevada Geothermal Power
01/02/07	GS aquires Texas energy company TXU	Bloomberg
01/02/07	Standard&Poor's will acquire the market leading Goldman Sachs Commodity Index	Bloomberg
01/03/07	Goldman Sachs to buy into South Korean alternative energy company	Market Watch
01/06/07	Morgan Stanley to supply crude oil to INEOS	Reuters
01/08/07	MS creates MS Carbon Bank	Carbon Insider
01/11/07	Goldman Sachs Plugs Into India Power-Trading Deal	Money Morning
01/01/08	Morgan Stanley Acquires Stake In Clean Technology Venture Investor NGEN Partners, LLC	Business Service Industry
01/02/08	GS $14 million investment in APX	Aletho News
01/08/08	GS & MS part of consortium aquiring 20% stake in DME	DME
01/09/08	MS will be largest sharehold of tidal energy company, Atlantis Resources Corporation	The Oil Voice
01/09/08	Morgan Stanley withdraws from Platts oil trading	The Times
01/10/08	GS purchases majority of E+Co, a carbon offset company	Greeninc Nytimes Blog
01/10/08	Goldman faces limits in Platts oil window-sources	Forbes
01/11/08	GS invested in APX, Carbon Offsets	Greeninc NYtimes Blog
01/12/08	Goldman Sachs Buys Into Carbon Offsets	Greeninc NYtimes Blog
01/12/08	Morgan Stanley Private Equity acquires minority stake in Biotor India	Greeninc NYtimes Blog
01/01/09	Goldman Sahs Buys Constellation Energy's Carbon Trading Operation	Climate Invest
01/01/09	Did Goldmans Goose Oil?	Forbes
01/02/09	Goldman Sachs Sudsidiary Buys Mojave Solar Plants	Sustainable Business
01/02/09	Goldman Sachs power arm buys California solar plants	Reuters
01/03/09	MS tripled their fuel storage capacity to over 1.5 million barrels in the Middle East	Reuters
01/06/09	MS acquires Triana, a private oil and gas exploration company	AlacraStore
01/07/09	Morgan Stanley invests 100 million to finance wind farm project	Reuters

References

Blanchard, O.J., Riggi, M. 2009. Why are the 2000s so different from the 1970s? A structural interpretation of changes in the macroeconomic effects of oil prices. *NBER Working Paper* #15467.

Büyükşahin, B., Haigh, M.S., Harris, J.H., Overdahl, J.A., and Robe, M.A. 2008. Fundamentals, Trader Activity and Derivative Pricing. *SSRN Working Paper* #966692.

Caballero, R., Farhi, E., Gourinchas, P. 2008, Financial Crash, Commodity Prices and Global Imbalances. *Brookings Papers on Economic Activity* (2), 1-55.

Cabinet Office. 2008. The rise and fall in oil prices: analysis of fundamental and financial drivers. *UK Cabinet Office*, Global Energy Team.

Chevalier, J.M., Laffitte, M., Baule, F., Lasserre, F., Odonnat, I., Viellefond, E. 2010. Report of the Working Group on the Volatility of Oil Prices. *French Ministry of Economics, Industry and Employment*, available (in French) at `http://www.minefe.gouv.fr/services/rap10/100211 chevalier-report-eng.pdf`.

Cifarelli, G., Paladino, G. 2010. Oil price dynamics and speculation: a multivariate financial approach. *Energy Economics* 32(2), 363-372.

CFTC, 2008. Staff Report on Commodity Swap Dealers and Index Traders with Commission Recommendations. *Commodities Futures Trading Commission*, Washington D.C., USA.

Cooper, J.C.B. 2003. Price Elasticity of Demand for Crude Oil: Estimates for 23 Countries. *OPEC Review* 27(1), 1-8.

Dvir, E., Rogoff, K.S. 2010. The Three Epochs of Oil. *Working Paper*, Harvard University, USA.

Hamilton, J.D. 2008. Understanding Crude Oil Prices. *Working Paper*, University of California, San Diego, USA.

Hamilton, J.D. 2009. Causes and Consequences of the Oil Shock of 2007-08. *Brookings Papers on Economic Activity* 1, 215-261.

Irwin, S.H., Sanders, D.R., Merrin, R.P. 2009. Devil or Angel? The Role of Speculation in the Recent Commodity Price Boom (and Bust). *Journal of Agricultural and Applied Economics* 41(2), 377-391.

Kaufmann, R.K., Ullman, B. 2009. Oil prices, speculation, and fundamentals: Interpreting causal relations among spot and futures prices. *Energy Economics* 31, 550-558.

Kilian, L. 2008a. Exogenous Oil Supply Shocks: How Big Are They and How Much Do They Matter for the U.S. Economy? *The Review of Economics and Statistics* 90(2), 216-240.

Kilian, L. 2008b. A Comparison of the Effects of Exogenous Oil Supply Shocks on Output and Inflation in the G7 Countries. *Journal of the European Economic Association* 6(1), 78-121.

Kilian, L. 2009. Not All Oil Price Shocks Are Alike: Disentangling Demand and Supply Shocks in the Crude Oil Market. *The American Economic Review* 99(3), 1053-1069.

Kilian, L. 2010. Explaining Fluctuations in Gasoline Prices: A Joint Model of the Global Crude Oil Market and the U.S. Retail Gasoline Market. *The Energy Journal* 31(2), 105-130.

Medlock III, K.B., Jaffe, A.M. 2009. Who is in the oil futures market and how has it changed? *Baker Institute Study*, Rice University, USA.

Parsons, J.E. 2009. Black Gold & Fool's Gold: Speculation in the Oil Futures Market. *MIT CEEPR Working Paper* #09-013.

Sanders, D.R., Boris, K., Manfredo, M. 2004. Hedgers, funds, and small speculators in the energy futures markets: an analysis of the CFTC's Commitments of Traders reports. *Energy Economics* 26, 425-445.

Saporta, V., Trott, M., Tudela, M. 2009. What can be said about the rise and fall in oil prices? *Bank of England Quarterly Bulletin* 49(3), 215-225.

Sornette, D., Woodard, R., Zhou, W.X. 2009. The 2006-2008 oil bubble: Evidence of speculation, and prediction. *Physica A: Statistical Mechanics and its Applications* 388(8), 1571-1576.

Tang, K., Xiong, W. 2010. Index Investing and the Financialization of Commodities. *Working Paper*, Princeton University, USA.

Till, H. 2009. Has There Been Excessive Speculation in the US Oil Futures Markets? What Can We (Carefully) Conclude from New CFTC Data? *Working Paper*. EDHEC Risk Institute.

Working, H. 1960. Speculation on Hedging Markets. *Stanford University Food Research Institute Studies* 1(2), 185-220.

Zagaglia, P. Macroeconomic factors and oil futures prices: A data-rich model. *Energy Economics* 32, 409-417.

In: Crude Oils
Editor: Claire Valenti

ISBN: 978-1-63117-950-1

Chapter 4

GLYCERIN FROM BIODIESEL IN POULTRY NUTRITION AS AN ALTERNATIVE FOR REDUCING THE ENVIRONMENTAL IMPACT

Verônica Maria Pereira Bernardino*[*,1], *Luciana de Paula Naves*[†,1] and *Carolina Fontes Prezotto*[‡,2]

[1]Federal University of Lavras – UFLA, Lavras, Minas Gerais, Brazil
[2]Federal University of Minas Gerais – UFMG, Belo Horizonte, Minas Gerais, Brazil

ABSTRACT

Fuel production from renewable resources has been intensified worldwide. Biofuels have the advantage of causing less environmental pollution than petroleum fuels. Brazil stands among the countries with the largest biodiesel production, 2.72 billion liters in 2012, according to recent data from the National Agency of Petroleum, Natural Gas and Biofuels (ANP). For each liter of biodiesel produced 100 ml of crude glycerin is generated, therefore Brazil alone had a production of 272

* Email: veronicampb@gmail.com; PhD in Non-ruminant Nutrition.
† Email: luciana.naves@hotmail.com; Postdoc in Non-ruminant Nutrition.
‡ Email: carolpre@yahoo.com.br; PhD student in Animal Science.

million liters of crude glycerin. As a global consequence, we currently have a glycerin production above the market demand, which requires researches in order to ensure new ways of using this co-product. In addition to enabling an extra profit in the supply chain, defining new alternatives to using glycerin is important to minimize an environmental obstacle, as to date there is still no well-defined law on how the discharge of excess glycerin must occur. Poultry industry is a highly developed economic activity in Brazil and in other countries such as the U.S.A., China and EU countries. Another important fact is that these countries also highlighted the global biodiesel production. Thus poultry production has been recognized as an interesting alternative to increase the demand of glycerin. Several studies have evaluated the use of crude or processed glycerin in animal feed and showed that glycerin can be considered a good source of dietary energy for poultry, also representing an opportunity that fits the need of producing quality meat with environmental responsibility. During the discussion in this chapter the results of the most significant studies on glycerin use in poultry nutrition published in high impact journals were considered. To date, it is not possible yet to define a single glycerin inclusion level in poultry diet that is properly applicable to all the different situations in animal production. Overall, the compilation of the work contemplated in this review suggests that the safe rate of glycerin inclusion is 5% for broilers, 7.5% for laying hens and from 4 to 8% for quails.

1. INTRODUCTION

The demand for energy required by the world population is increasing and, on the other hand, fossil-fuels are not only expensive but also obtained from non-renewable natural resources that have the ability to cause a high environmental impact (Boso et al., 2013). Therefore, biofuels represent an important alternative to replacing petroleum fuels.

Crude glycerin is a byproduct of biodiesel production obtained from transesterification reactions, that is, reactions between lipids (oils and/or fats) and an alcohol in the presence of a catalyst (Van Gerpen, 2005). Approximately 10% of the total volume of biodiesel produced corresponds to glycerine (Dasari et al., 2005). Thus, the increase in biodiesel production in recent years has resulted in a supply of crude glycerin higher than the capacity of utilization by the chemical market. Moreover, to date, there is no legislation to regulate the proper disposal of the exceeding glycerin, which can lead to serious environmental problems if all the glycerin produced is not used. In this context, a major challenge today is the development of technologies that enable new ways for using glycerin, avoiding their disposal in the environment and providing new ways to make a profit with this co-product.

In poultry production, feed represents about 70% of the production cost and energy is one of the most expensive formulation component (Bertechini, 2012), which justifies the interest of nutritionists in evaluating alternative energy foods that can replace conventional food efficiently without harming the animals performance (Batista et al., 2013). Therefore, glycerin has been increasingly studied as a possible energetic ingredient in feed for broilers, hens and quails (Cerrate et al. 2006; Dozier et al. 2008; Swiatkiewicz and Koreleski, 2009; Silva et al., 2012; Silva et al., 2013). It is also important to consider that there is the expectation that the increased rate of compulsory inclusion of biodiesel to petroleum diesel will increase the supply of glycerin in the market, encouraging the price reduction of this alternative food culminating in decreased poultry production costs when using glycerin in poultry feed.

In Brazil the use of glycerin was permitted by Resolution No. 386 of August 5, 1999 (Lopes et al., 2012). In the United States and Europe, glycerin has applications in food industry as a food additive and has GRAS status, that is, is recognized as a safe food (Menten et al., 2010). However, in order to avoid cases of animal poisoning, problems with high moisture content in excreta and trying to standardize the composition of the glycerides produced, the Ministry of Agriculture, Livestock and Supply (MAPA, Brazil) determined that glycerin added as an ingredient in poultry feed should have a maximum of 150 mg/kg methanol and 13% moisture content, having at least 80% glycerol and minimum sodium or other electrolyte possible(Lopes et al., 2012).

Therefore it is very important to ensure that the inclusion level does not exceed the metabolic capacity of the organism, because if this occurs, there may be losses in productive performance (Guerra et al., 2011; McLea et al., 2011; Jung and Batal, 2011; Silva et al. 2013) together with other problems such as increased water excretion, which is undesirable because it requires a more laborious handling during the poultry production (Gianfelici et al., 2011). Thus, during the discussion presented in this chapter, results of the most important papers regarding glycerine use in poultry nutrition published in scientific journals, including classics from the early studies to the latest research will be considered.

2. Worldwide Biodiesel and Glycerin Production

Biodiesel is a fuel derived from renewable resources that has emerged as an alternative to fossil fuels, especially because it is less polluting than petroleum fuels (Boyle, 1998). The raw materials for biodiesel production can

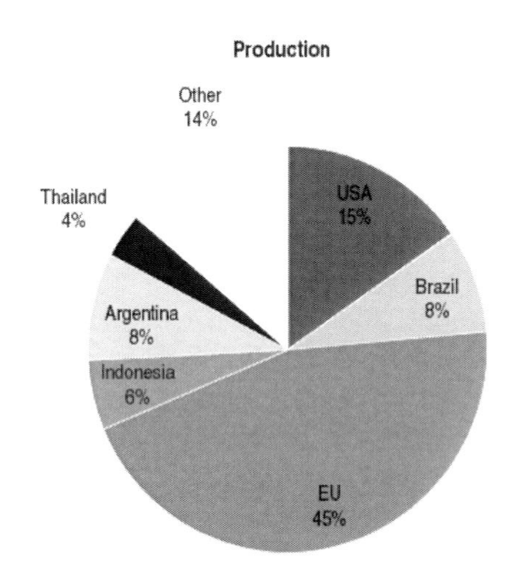

Source: OECD and FAO Secretariats.

Use

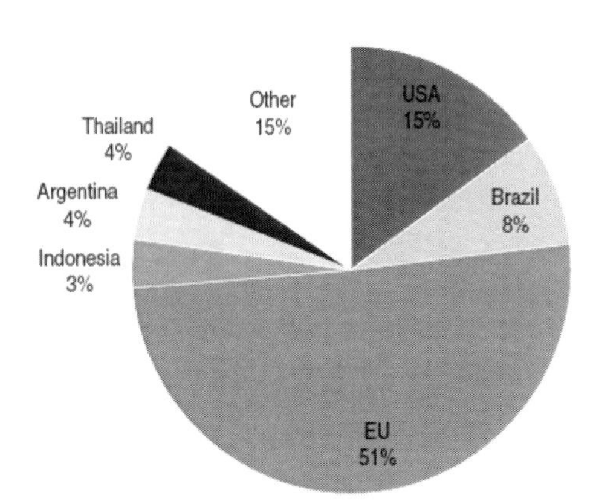

Figure 1. Regional distribution of world biodiesel production and use in 2022.

be vegetable oils (cottonseed, groundnut, babassu, canola, dende, sunflower, castor bean, soybean, etc.), animal fats (beef tallow, fish oil, lard), besides residual oils and fats derived from domestic, commercial and industrial processing (Garcia and Tookuni, 2006; ANP, 2013).

In Brazil, Law 11.097/05 established that from January/2008 all diesel sold in the country should contain 2% biodiesel. However, in 2010, the Resolution No. 6/2009 of the National Energy Policy Council (CNPE) came into effect determining the mandatory inclusion of 5% biodiesel to petroleum diesel. In Brazil, the biodiesel production registered in 2012 was 2.72 billion liters (Garcia and Tookuni, 2006; ANP, 2013).

The biodiesel production is growing every year not only in Brazil but worldwide, with an expected world production of approximately 41 bnl by 2022. EU countries, USA, Brazil and Argentina stand out among the main biodiesel-producing and consuming regions (OECD-FAO, 2013) (Figure 1).

Biodiesel is obtained by vegetable oil and/or animal fat transesterification, and this production process generates crude glycerine as a co-product (Ma & Hanna, 1999; Van Gerpen, 2005). The basic substances such as sodium and potassium hydroxides are frequently used as catalysts in the transesterification (Thompson and He, 2006). The raw material type and the biodiesel production process influence the composition and the quality that the crude glycerin will display. Furthermore, crude glycerin can undergo some degree of processing or purification, which can also affect its final quality and chemical composition. Therefore, there are several types of glycerins available in the market, and it is necessary to assess whether distinct glycerins can provide different zootechnical results when they are supplied in bird feeds.

3. Glycerin in Poultry Nutrition

3.1. Glycerin in Diets for Broilers

Glycerin has been a good energetic ingredient in broiler diets. The following energy values have been determined for glycerins used in these diets: 3.527 kcal apparent metabolizable energy – AME/kg (Cerrate et al., 2006), 3.434 kcal nitrogen-corrected apparent metabolizable energy – AMEn/kg (Dozier et al., 2008), 3.500 kcal AME/kg (Fernandes et al., 2010), from 3.254 to 4.134 kcal AMEn/kg (Dozier et al., 2011), 3.276 kcal AMEn/kg (Gianfelici et al., 2011), 3.422 kcal AME/kg (Silva et al., 2012), 3.069 kcal

AMEn/kg (Batista et al., 2013), from 3.595 to 4.908 kcal AMEn/kg (Lima et al., 2013).

Cerrate et al. (2006) evaluated the effect of diets with 2.5, 5 and 10% crude glycerin on the performance, carcass yield and cuts of a commercial broiler strain (Cobb 500) during 1-42 days of age, and concluded that the inclusion of up to 5% glycerin did not affect the performance and carcass yield, and was also able to increase the birds breast yield. However, they found that the inclusion of 10% glycerol reduced performance and increased the moisture content of excreta. Similarly, Silva et al. (2012) also evaluated increasing levels (2.5, 5.0, 7.5 and 10%) of crude glycerin from soybean oil in diets for broilers in the period from 1 to 42 days of age, and concluded that although feed conversion in poultry has not been changed, glycerin inclusion levels above 5% increases litter humidity.

Cerrate et al. (2006) attributed the increase in the excreta moisture content as a result of a diet with 10% glycerin containing excess potassium (0.15% more than the control diet, formulated without glycerin) which probably caused an electrolyte imbalance in the feed. The high potassium content in the diet with 10% glycerin was associated with residual potassium levels from using potassium hydroxide as catalyst in the biodiesel production.Also in order to maintain the feed isoproteic, it was necessary to add more soybean meal (source of potassium) in the diets with a greater glycerin inclusion, which may also have contributed with a wetter excreta. In addition, the excess of glycerol excreted may give a wet litter appearance since this compound has a hygroscopic characteristic and aggregates water when excreted.

In the same experiment previously mentioned, Cerrate et al. (2006) also found that the water incorporation in the carcass was not affected by glycerin inclusion, disagreeing with Simon (1996) who suggested that glycerol could promote beneficial effects on broiler carcass water intake because of its easy retention in chicken muscle, probably due to an increase in blood osmolality (Riedesel et al., 1987; Montner et al., 1999; Wingo et al., 2004).

According to Lopes et al. (2012), it is possible to include 10% glycerin in starter diet, using only up to 5% glycerin inclusion in other production phases, showing that the bird age has influence on the use of glycerin from the diet. Recently, Lima et al. (2013) determined the AMEn of three different glycerins in Cobb 500 male broilers at different ages (1 to 10, 11 to 20, 21 to 30 and 31 to 40 days of age). The AMEn average values determined were (in dry matter): 3,595 kcal/kg for crude glycerine from soybean oil, 4,908 kcal/kg for mixed crude glycerine from frying oil and lard, and 3,774 kcal/kg for semipurified glycerine from soybean oil. Furthermore, there was a linear decrease in AMEn values with the increasing age of the broilers, in other words, younger broilers have a higher capacity of energy utilization from these feedstuffs. However, Lima et al. (2013) reported that more studies regarding the enzymatic action or metabolism of nutrients contained in the glycerine products must be done to elucidate the differences between young and old broilers.

Dozier et al. (2011) evaluated 5 different types of glicerin provided for Ross x Ross 708 male broilers, in the period from 17 to 22 days of age. Crude glycerin samples were obtained from various biodiesel plants that used soybean oil (hexane and extruded soybean), tallow, yellow grease, or poultry fat. For each type of glycerin, a control diet without glycerin and a diet containing 6% of this ingredient were evaluated. The assessed AMEn from crude glycerin ranged from 3254 to 4134 kcal / kg, demonstrating that crude glycerin is a good energetic ingredient for broilers; however, its energy value can be influenced by its chemical composition, which in turn, is influenced predominantly by their raw materials and by the processing used during the biodiesel production. Furthermore, Dozier et al. (2011) proposed the following equation to predict the crude glycerin AMEn based on its chemical composition: AMEn (kcal / kg) = 1.605 - (19.13 x percentage% of methanol) + (39.06 x percentage% of fatty acids) + (23.47 x percentage% of glycerin).

The crude glycerin energy value from soybean oil was also evaluated by Gianfelici et al. (2011) in an experiment with Ross 308 male broilers in the period from 35 to 42 days of age.The experimental diets were formulated with increasing levels of glycerin inclusion (0, 5, 10, 15 and 20%) and it was observed that the evaluated glycerin AMEn average value was 3,561 kcal/kg for dry matter 3,276 kcal / kg for natural matter. Furthermore, Gianfelici et al. (2011) found the highest AMEn value (4,890 kcal/kg for dry matter) obtained with a 15% glycerin rate; however this level can not be recommended because it results in high moisture levels in the birds excreta (causing clinical cases of diarrhea). As explanations for this result, the researchers suggested that it is likely that the highest levels of glycerin evaluated have exceeded the broilers glycerol metabolism capacity (eg, caused by a saturation of the glycerol kinase activity).Thus, the glycerol adsorbed, but not metabolized, requires its excretion in the urine. However, as glycerol is water soluble, its presence in urine increases the concentration of water in the birds' excreta. Another factor that may have contributed is the glycerin's high level of sodium (2.05%).

Kim et al. (2013) observed that the inclusion of 5% crude glycerin in the Ross x Ross 308 male broilers diet resulted in bowel transit time of the digesta similar to that determined for birds fed with a control diet without glycerin (119 versus 112 minutes, respectively). The reason for this observation may be that crude glycerine used in this experiment contained very small amounts of free fatty acids or other fat-derivatives (e.g., monoglycerides), which are known as the primary regulators of intestinal hormones and motility (Hildebrand et al., 1998; Little et al., 2007) to influence the intestinal transit time. Moreover, the diet containing 5.0% crude glycerine had greater apparent total tract digestibility (ATTD) of dry matter and gross energy than the basal diet, but no differences were observed for other evaluated nutrients (ATTD of CP, crude fat, crude ash, and phosphorus). In conclusion, it is likely, therefore, that the effect of dietary supplementation of crude glycerine on the ATTD of dry matter and gross energy observed in this experiment was caused by different ATTD of crude glycerine itself from the basal diet containing corn and soybean meal rather than by the change in intestinal transit time.

An experiment with Cobb-500 male broilers was conducted in other to determine the maximum crude glycerine level (a primary by-product of biodiesel production) considering maintaining the broilers performance. For this, Jung and Batal (2011) proposed an experimental design containing various combinations of experimental diets based on corn and soybean meal, so that the concentrations of crude glycerin in diets ranged from 0 to 10% inclusion, while methanol concentration in the feed ranged from 0,01 to 3,10%. They concluded that the broilers performance was not affected by addition of up to 5% crude glycerin regardless the final concentration of methanol in the feed (from 0,01 to 3.10%).

The crude glycerin mixture obtained from biodiesel production with raw materials from plant and animal origin has also been evaluated in Cobb broilers feeding in the rearing period from 1 to 42 days of age (Guerra et al., 2011). In that experiment the results showed that the use of up to 10% crude glycerin in formulated feed based on corn and soybean meal did not affect carcass yield and cuts, nor the body chemical composition at 21 and at 42 days of age. These results corroborate with Simon et al. (1996) who assessed the performance of broiler chicks fed diets containing 0, 5, 10 and 25% glycerin and concluded that up to 10% of this product may be included in the diet without affecting the animals'performance. However, according to Guerra et al. (2011), the maximum level of inclusion should be 5% because in higher concentrations occurs a reduction in birds zootechnical performance (feed intake, weight gain and body weight) and an increase in moisture bedding content may occur.

On the other hand McLea et al. (2011) conducted an experiment with Ross broiler chickens, within 7 to 28 days of age, to evaluate the optimal crude glycerin inclusion level in diets based on wheat and soybean meal.The study was designed as a 3 x 2 + 1 factorial design with 3 concentrations (3.3, 6.7 and 10.0%) of glycerine from 2 sources, A and B and a control diet. They concluded that the glycerine source did not affect performance and that increasing level of glycerol improved feed conversion ratio, with 6.7% of inclusion resulting in the most efficient conversion of feed to gain without any negative effects upon nutrient digestibility, as dry matter digestibility, glycerol digestibility, total starch digestibility, AME and AME:gain.

It is important to know to what level the inclusion of glycerin can be metabolized by birds. Therefore, three different sources of glycerin (cCrude glycerine from soybean oil – CGSO, mixed crude glycerine from frying oil and lard – MCG and lardand semipurified glycerine from soybean oil – SPGSO) were evaluated in diets for male Cobb-500 broilers, at the age of 22 to 35 and 33 to 42 days (Bernardino et al., 2013b). Two independent experiments were performed, and each type of glycerine was evaluated at four concentrations in the diet (1.75, 3.5, 5.25 and 7.0% inclusion) for each period of rearing. For both experiments, increasing the concentration of glycerine led to an increase in the glycerol kinase activity and in the plasmatic glycerol concentration indicating that the glycerol present in the glycerine is absorbed by the intestine and can be transported to the liver through the bloodstream. The increase in the activity of this enzyme demonstrates physiological self-regulation by the organism to allow better use of the glycerol from the diet; this supports the use of glycerine in the nutrition of broilers, because the glycerol-3-phosphate produced by glycerol kinase can be converted to dihydroxyacetone-phosphate (DHAP) by the catalytic action of another enzyme, glycerol-3-phosphate dehydrogenase, also present in the liver. The DHAP can be converted to glyceraldehyde-3-phosphate by triosephosphateisomerase that is then used metabolically for the synthesis of glucose (gluconeogenesis), fatty acids (lipogenesis) or is completely oxidized for the production of energy via glucolysis and the Krebs Cycle (Champe et al., 2009). Therefore, Bernardino et al. (2013b) concluded that, based on the absence of saturation of the glycerol kinase activity for the three glycerine sources and for both rearing periods evaluated, the broilers can metabolize the glycerol (at the level of the phosphorylation of the glycerol to glycerol-3-phosphate in the liver) present in the glycerine when the diet is supplemented with up to 7.0% of glycerine.

In another experiment, Bernardino et al. (2013a) evaluated whether the addition of CGSO, MCG and SPGSO at inclusion levels of 1.75, 3.5, 5.25 and 7.0%, affects the activity of hepatic glutamate dehydrogenase, performance and protein content in the breast of broilers Cobb-500, during 22 to 35 and 33 to 42 days of age. This hypothesis was tested since according to Steele et al. (1971), glycerol in the diet can spare glucogenic amino acids by inhibiting the activity of enzymes, such as glutamate dehydrogenase, thereby promoting protein deposition in muscle tissues. However, they concluded for both rearing periods evaluated, that an increase in glycerine in the diet did not necessarily reduce the activity of hepatic glutamate dehydrogenase; in addition, the protein deposition in the breast of broilers is not strictly correlated with the activity of this enzyme. Furthermore, in a previous scientific publication, Bernardino et al. (2012) reported that the inclusion of up to 7% of any of the three evaluated glycerins did not alter the broilers plasma creatinine concentration in the period from 33 to 42 days of age, relating this result to a possible absence of kidney injury in birds.

3.2. Glycerin in Diets for Laying Hens

The metabolizable energy value of the glycerin from biodiesel is variable, dependent on the glycerol and fatty acids content in these coproducts and the species and age of the animal studied, among other factors (Dozier et al., 2008). Glycerin has also shown to be a good energetic ingredient in laying hens diet, with AMEn mean values ranging from 3,805 kcal (Lammers et al., 2008) and 3,970 kcal of AMEn/kg (Świątkiewicz and Koreleski, 2009). Świątkiewicz and Koreleski (2009) evaluated the effect of adding 0, 2, 4 and 6% crude glycerin (a coproduct of commercial biodiesel production from rapeseed) in Bovans Brown laying hens diet, with 28-53 weeks of age. They concluded that the crude glycerin may be incorporated to a level of 6% in the diet of laying hens without any detrimental effect on egg performance parameters (egg production, egg weight, daily egg mass, daily feed consumption and feed conversion), egg quality parameters (albúmen height, Haugh units, yolk color and thickness, density and breaking strength of eggshell), nutrient retention (N, Ca and P) and metabolizability of energy.

Slightly higher glycerin inclusion levels were evaluated by Boso et al. (2013). The crude glycerin used was resultant from the biodiesel production with soybean oil and was supplemented in Hy-Line W36 laying hens' diet at inclusion levels of 0, 1.5, 3.0, 4.5, 6.0 and 7. 5%. The experiment lasted 112 days (divided into 4 cycles of 28 days each) and started when the hens were 35 weeks old. The parameters of feed intake, feed conversion, excreta moisture content, specific gravity, Haugh unit, shell thickness and percentage of eggs were not affected by the glycerin level in the diet. Furthermore, these researchers noted that increasing the glycerin in the feed linearly increased egg production as well as the deposition of fatty acids (FAs) in eggs (more specifically linoleic and behenic fatty acids and representatives of the omega 6 series). This improvement in the lipid profile was explained by the high content of evaluated glycerin polyunsaturated FAs and because the birds have a high absorption coefficient of these lipids.It is noteworthy that the enrichment of eggs with polyunsaturated FAs is desirable because these molecules promote human health by reducing the incidence of diseases such as cancer and diabetes (Woods and Fearon, 2009). Therefore, according Boso et al. (2013), crude glycerin from soybean oil can be included in laying hens feed at a concentration of up to 7.5% providing improvements in the chickens growth performance, in the fatty acid profile and quality of the eggs, besides it does not influence the excreta moisture content.

A glycerin from the biodiesel production with soybean oil was also included (0, 2.5, 5.0 and 7.5% glycerin) in diets for Lohman Brown laying hens, with 39-55 weeks of age (Yalçın et al., 2010).The dietary treatments did not significantly affect body weight, egg production, egg weight, feed efficiency, mortality, egg albumen index, egg yolk index and egg Haugh unit, yolk weight percentage, exterior egg quality characteristics, excreta moisture, heterophils to lymphocytes ratio (H/L ratio), blood parameters, antibody production to SRBC and excreta moisture. Therefore the glycerine can be used at 7.5% in the diets of laying hens without adverse effects on the measured parameters.

3.3. Glycerin in Diets for Quail

The use of glycerin in diets for quail has also been evaluated, although less often than for other birds such as broilers, so there is the need for further studies in order to define the optimal level of glycerin inclusion in these animals' diets.

Gasparino et al. (2012) evaluated Japanese quails fed with increasing crude glycerine diet levels (0, 4 and 8%) at 0 to 35 days of age. The effects on performance, expression of mRNA of growth hormone (GH) and insulin growth factor (IGF-1) were evaluated. The authors observed that supplementation of 8% glycerin reduced the mRNA expression of GH and IGF-1 in the breast of the birds, in comparison to results determined for quails fed with control diet (without glycerin) or diet containing 4 % crude glycerin. Furthermore, the use of 8% glycerin in the diet decreased the feed conversion.

Therefore, Gasparino et al. (2012) suggested that crude glycerin can be included up to 4% in feed without harming the birds' performance and the expression rate of the genes evaluated.

Silva et al. (2013) evaluated the effect of adding crude glycerin in the diet (at inclusion levels of 0, 8 and 12%) on the performance of Japanese quails (*Coturnixcoturnix japonica*) with 1-28 days of age.The results showed that the inclusion of 8% glycerin did not impair the performance of the birds, however, the addition of crude glycerin at the level of 12% increased feed intake and decreased feed conversion. According to Silva et al. (2013), the increased feed intake may have occurred due to a greater palatability of the diet containing 12% glycerin, due to the sweet taste of glycerol. Consequently, feed conversion decreased because the weight gain was not changed at any level of the glycerin evaluated.

Pasquetti (2011) conducted two experiments with meat type quail, between 1-14 and 15-35 days of age, evaluating the effects of increasing levels of a crude glycerin and a semi-purified glycerin (3, 6, 9, 0:15%) on the birds'performance. In the period between 1 and 14 days Pasquetti (2011) reported that the two glycerides can be used in the quail feeding at inclusion rate up to 10%. As for the period from 15 to 35 days of age, the addition of glycerin may be up to 15%. Similarly, recently, Batista et al. (2013) conducted three experiments with quail (*Coturnix coturnix sp*) fed diets containing semipurified vegetable glycerin. It was concluded at the end of experiment I that the values for gross energy and metabolizable energy corrected for nitrogen balance of glycerin were 3,585 and 3,069 kcal/kg for fresh matter, respectively. In the other experiments, the performance parameters were evaluated in the periods from 1 to 14 (experiment II) and 15 to 35 days of age (experiment III) after feeding the quail diets with increasing inclusion levels of glycerin (0; 4, 8, 12 and 16%). Those researchers reported that an increase in the level of glycerin inclusion (from 1 to 14 days) gradually decreased the feed conversion, ranging from 1.58 g / g (4% glycerin) up to 1.85 g / g (16% glycerin). However, in the period from 15 to 35 days of age, the semipurified glycerin can be used as an energy source in feed up the level of 16% inclusion without affecting the quail's performance.

Erol et al. (2009) evaluated the inclusion of up to 10% crude glycerin in diets on the laying quails' performance during an 18-week period. The glycerin AMEn was of 3,350 kcal / kg, and the glycerin contained 90.2% glycerol. Only the inclusion of 10% glycerin worsened the egg quality, lowering the albumen height, albumin and Haugh unit, and increased the cholesterol content in the yolk. Those authors concluded that crude glycerin can also be considered an energetic ingredient for laying quails and could be included in the diet at inclusion rates of up to 7.5% without impairing the productive performance and egg quality.

4. Final Remarks

This literature review found that several factors seem to influence what will be the maximum level of glycerin inclusion, highlighting the following factors: species, age and lineage of birds; glycerin composition (which, in turn, is influenced predominantly by raw material, the processing used for biodiesel production and the degree of processing applied to the crude glycerin produced , if any); as well as plant ingredients which predominate in the diet (corn and soybean meal or wheat and soybean meal, for example). Therefore, it is not possible yet to define a single glycerin inclusion level in poultry diet that is properly applicable to all the different situations of animal production. Overall, the compilation of the work contemplated in this review suggests that the safe rate of inclusion of glycerin is 5% for broilers, 7.5% for laying hens and from 4 to 8% for quails.

It is important to emphasize the need for future studies on the economic feasibility of glycerin in poultry nutrition, contemplating an important aspect that should be considered to demonstrate the real effectiveness or not of including this co-product in poultry feed. However, in our present situation, this is not an easy task due to many factors such as the diversity of glycerin produced from biodiesel and variation in glycerin costs according to its origin, degree of processing and amount of biodiesel produced.Therefore, the first steps to allow further economic analysis of the economic viability of glycerin include price standardization and glycerin chemical composition as well as the establishment of reliable average energy values onon calculations of economic feasibility and other factors correlated.

References

ANP – National Agency of Petroleum, Natural Gas and Biofuels (Brazil). 2013: Available: http://www.anp.gov.br. Accessed December 07, 2013.

Batista, E.; Furlan, A.C.; Ton, A.P.S.; Pasquetti, T.J.; Quadros, T.C.O.; Grieser, D.O.; Zancanela, V., 2013: Nutritional evaluation of vegetable semi-purified glycerin of meat type quail. *Arquivo Brasileiro de Medicina Veterinária e Zootecnia* 65, 1783–1791.

Bernardino, V.M.P.; Rodrigues, P.B.; Prezotto, C.F.; Oliveira, D.H.; Teixeira, L. do V.; Nardelli, N.B.S., 2012: *Creatinine in the plasma of broiler fed diets containing different sources and levels of glycerine.* In: 49ª Annual Meeting of the Brazilian Society of Animal Science. Brasil, p. 1–4.

Bernardino, V.M.P.; Rodrigues, P.B.; Naves, L. de P.; Zangeronimo, M.G.; Alvarenga, R.R.; Rosa, P.V.; Santos, L.M. dos; Teixeira, L. do V., 2013a: Activity of glutamate dehydrogenase and protein content in the breast of broilers fed diets containing different sources and levels of glycerine. *Journal of Animal Physiology and Animal Nutrition,* Article first published online: 3 AUG 2013, DOI: 10.1111/jpn.12113.

Bernardino, V.M.P.; Rodrigues, P.B.; Naves, L. de P.; Rosa, P.V.; Zangeronimo, M.G.; Gomide, E.M.; Saldanha, M.M.; Alvarenga, R.R., 2013b: Content of plasmatic glycerol and activity of hepatic glycerol kinase in broiler chickens fed diets containing different sources and concentrations of glycerine. *Journal of Animal Physiology and Animal Nutrition,* Article first published online: 21 MAY 2013, DOI: 10.1111/jpn.12083.

Bertechini, A.G., 2012: *Monogastric nutrition.* Editora UFLA : Universidade Federal de Lavras, Brazil. 373p.

Boso, K.M.O.; Murakami, A.E.; Duarte, C.R.A.; Nascimento, G.R.; Matumoto-Pintro, P.T.; Ospina-Rojas, I.C., 2013: Fatty acid profile, performance and quality of eggs from laying hens fed with crude vegetable glycerine. *International Journal of Poultry Science*12, 341–347.

Boyle, G., 1998: Renewable energy: power for a sustainable future. New York: *Oxford University Press.*

Cerrate, S.; Yan, F.; Wang, Z., Coto, C., Sacakli, P., Waldroup, P.W., 2006: Evaluation ofglycerine from biodiesel production as a feed ingredient for broilers. *International Journal of Proultry Science*5, 1001–1007.

Champe, P. C.; Harvey, R. A.; Ferrier, D. R., 2009: *Biochemistry: Lippincott's illustrated Reviews*, 4th edn. Artmed, Porto Alegre, 528 p.

Dasari, M.A.; Kiatsimkul, P.P.; Sutterlin, W.R., 2005: Low-pressure hydrogenolysis of glycerol to propylene glycol. *Applied Catalysis. A: General* 281, 225–231.

Dozier, W.A.; Kerr, B.J.; Corzo, A.; Kidd, M.T.; Weber, T.E.; Bregendah, K., 2008: Apparent metabolizable energy of glycerin for broiler chickens. *Poultry Science* 87, 317–322.

Dozier, W.A.; Kerr, B.J.; Branton, S.L., 2011: Apparent metabolizable energy of crude glycerin originating from different sources in broiler chickens. *Poultry Science* 90, 2528–2534.

Erol, H.; Yalçin, S.; Midilli, M.; Yalçin, S., 2009: The effects of dietary glycerol on growth and laying performance, egg traits and some blood biochemical parameters in quails. *Revue de MedecineVeterinaire* 160, 469–476.

Fernandes, E.A; Machado, C.A.; Fagundes, N.S., França, A.M.S.; Ramos, G.C., 2010. Inclusion of purified glycerol in diets for broilers. In: *Conferência Apinco*, Santos, São Paulo, Brasil.

Garcia, A. J. M.; Tookuni, J. P. M., 2006: Biodiesel from animal fat. Biodiesel Magazine BR. Available: *www.biodieselbr.com/estudos/biodiesel/biodiesel-sebo-gordura-animal.htm.*

Gasparino, E.; Oliveira Neto, A.R.; Del Vesco, A.P.; Pires, A.V.; Batista, E.; Voltolini, D.M.; Souza, K.R.S., 2012: Expression of growth genes in response to glycerol use in Japanese quail diets. *Genetics and Molecular Research 11, 3063–3068.*

Gianfelici, M.F.; Ribeiro, A.M.L.; Penz Jr., A.M.; Kessler, A.M.; Vieira, M.M.; Machinsky, T., 2011: Determination of apparent metabolizable energy of crude glycerin in broilers chickens. *Brazilian Journal of Poultry Science* 13, 255–258.

Guerra, R.L. de H.; Murakami, A.E.; Garcia, A.F.Q.M.; Urgnani, F.J.; Moreira, I.; Picoli, K.P., 2011: Crude glycerine mixture in diets of broiler chickens (1 to 42 days). *Revista Brasileira de Saúde e Produção Animal* 12, 1038–1050.

Hildebrand, P.; Petrig, C.; Burckhardt, B.; Ketterer, S.; Lengsfeld, H.; Fleury, A.; Hadváry, P.; Beglinger, C., 1998: Hydrolysis of dietary fat by pancreatic lipase stimulates cholecystokinin release. *Gastroenterology* 114, 123–129.

Jung, B.; Batal, A.B., 2011: Nutritional and feeding value of crude glycerin for poultry. 2. Evaluation of feeding crude glycerin to broilers. *Journal of Applied Poultry Research* 20, 514–527.

Kim, J.H.; Seo, S.; Kim, C.H.; Kim, J.W.; Lee, B.B.; Lee, G.I.; Shin, H.S.; Kim, M.C.; Kil, D.Y., 2013: Effect of dietary supplementation of crude glycerol or tallow on intestinal transit time and utilization of energy and nutrients in diets fed to broiler chickens. *Livestock Science* 154, 165–168.

Lammers, P.J.; Kerr, B.J.; Honeyman, M.S..; Stalder, K.; Dozier, W.A.; Weber, T.E.; Kidd, M.T.; Bregendahl, K., 2008: Nitrogen-corrected apparent metabolizable energy value of crude glycerol for laying hens. *Poultry Science* 87, 104–107.

Lima, E.M.C.; Rodrigues, P.B.; Alvarenga, R.R.; Bernardino, V.M.P.; Makiyama, L.; Lima, R.R.; Cantarelli, V.S.; Zangeronimo, M.G., 2013: The energy value of biodiesel glycerine products fed to broilers at different ages. *Journal of Animal Physiology and Animal Nutrition* 97, 896–903.

Little, T.J.; Russo, A.; Meyer, J.H.; Horowitz, M.; Smyth, D.R.; Bellon, M.; Wishart, J.M.; Jones, K.L.; Feinle-Bisset, C., 2007: Free fatty acids have more potent effects on gastric emptying, gut hormones, and appetite than triacylglycerides. *Gastroenterology* 133, 1124–1131.

Lopes, M.; Pires, P.G. da S.; Nunes, J.K.; Roll, F.B.; Anciuti, M.A., 2012: Glycerin in the diet of broilers. *PUBVET* 6, 1–14.

MA, F.; Hanna, M. A.,1999: Biodiesel production: A review. *Bioresource Technology*70, 1–15.

McLea, L.; Ball, M.E.E.; Kilpatrick, D.; Elliott, C., 2011: The effect of glycerol inclusion on broiler performance and nutrient digestibility. *British Poultry Science* 52, 368–375.

Menten, J.F.M.; ZAVARIZE, K.C.; Silva, C.L.S. da, 2010. *Biodiesel: opportunities of the use of glycerine in poultry nutrition.* IV *Latin American Animal NutritionCongress* (CLANA). *Estância de São Pedro, São Paulo, Brazil: Colégio Brasileiro de Nutrição Anima*l. p. 43–56.

Montner, P.; Zou, Y.; Robergs, R.A.; Murata, G.; Stark, D.; Quinn, C.; Wood, S.; Lium, D.; Greene, E.R., 1999: Glycerol hiperhydration alters cardiovascular and renal function. *Journal of Exercise Physiology.* Available in: http://www.asep.org/asep/asep/jan12c.htm. Acessed in February, 07, 2014.

OECD-FAO Agricultural Outlook, 2013: *Chapter 3: Biofuels*, p.102–119, 2013.

Pasquetti, T.J., 2011: *Avaliação nutricional da glicerina bruta ou semipurificada, oriundas de gordura animal e óleo vegetal, para codornas de corte. Dissertação,* 107p., Universidade Estadual de Maringá, Maringá, Paraná, Brasil.

Riedesel, M.L.; Allen, D.Y.; Peake, G.T.; Al-Qattan, k., 1987: Hyperhydrationwithglycerolsolutions. *Journal Applied of Physiology* 51, 594–1600.

Silva, C.L.S.; Menten, J.F.M.; Traldi, A.B.; Pereira, R.; Zavarize, K.C.; Santarosa, J., 2012: Glycerine derived from biodiesel production as a feedstuff for broiler diets. *Brazilian Journal of Poultry Science*14, 193–202.

Silva, S.C.C. da; Gasparino, E.; Voltolini, D.M.; Marcato, S.M.; Tanamati, F., 2013: mRNA expression of mitochondrial genes and productive performance of quails fed with glycerol. *Pesquisa Agropecuária Brasileira*48, 228–233.

Simon, A.; Bergner, H.; Schwabe, M., 1996: Glycerol feed ingredient for broiler chickens. *Archives of Animal Nutrition* 49, 103–112.

Simon, A., 1996: Administration of glycerol to broilers in the drinking water. *Landbauforschung Volkenrode* 169, 168–170.

Steele, R.; Winkler, B.; Altszuler, N., 1971: Inhibition by infusion glycerol of gluconeogenesis from other precursors. *American Journal of Physiology* 221, 883–888.

Swiatkiewicz, S.; Koreleski, J., 2009: Effect of crude glycerin level in the diet of laying hens on egg performance and nutrient utilization. *Poultry Science* 48, 615–619.

Thompson, J.C.; He, B.B., 2006: Characterization of crude glycerol from biodiesel production from multiple feedstocks. *Applied Engineering in Agriculture* 22, 261–265.

Van Gerpen, J., 2005: Biodiesel processing and production. *Fuel Processing Technology* 86, 1097–1107.

Yalçın, S.; Erol, H.; Özsoy, B.; Onbaşılar, I.; Yalçın, S.; Üner, A., 2010: Effects of glycerol on performance, egg traits, some blood parameters and antibody production to SRBC of laying hens. *Livestock Science* 129, 129–134.

Wingo, J.E.; Casa, D.J.; Berger, E.M.; Dellis, W.O; Knight, J.C.; McClung, J.M., 2004: Influence of pre-exercise glycerol hydration beverage on performance and physiologic function during mountain-bike races in the heat. *Journal of Athletic Training* 39, 169–175.

Woods, V.B.; Fearon, A.M., 2009: Dietary sources of unsaturated fatty acids for animals and their transfer into meat, milk and eggs: A review. *Livestock Science* 126, 1–20.

In: Crude Oils
Editor: Claire Valenti
ISBN: 978-1-63117-950-1

Chapter 5

CRUDE OIL PRODUCTION: ITS ENVIRONMENTAL AND GLOBAL MARKET IMPACT*

Lawrence Atsegbua and Violet Aigbokhaevbo
The Faculty of Law, University of Benin, Nigeria

ABSTRACT

The primacy of crude oil as an energy source has provided a platform for social-economic development in most oil producing and consumer nations. Food, housing, transportation, investment services and inflationary trend is largely dependent on the volatility of crude oil prices. The anticipated shift in crude oil demand premised on the United States shale exploitation and the economic pandemonium generated in the Organization of Petroleum Exporting Countries (OPEC) has re-enacted crude oil pricing as the barometer for global economic trend. Crude oil exploitation, production, transportation and utilisation processes, however adversely impact on marine and coastal habitat. Pollution and infections attributable to toxic substances from crude oil and its derivatives can have long and short term health implications on animals and humans. Crude oil dependence is also contributory to global warming and its complications from climate change. This paper examines the environmental and global market impact of crude oil production, analyses attempts to curb inherent

---

* Prof. Lawrence Atsegbua (SAN) and Dr. Violet O. Aigbokhaevbo are both Lecturers in the Department of Public Law, University of Benin, Nigeria.

challenges through energy efficiency and the development of renewable alternative energy sources. It concludes that the political instability, insecurity, corruption and economic adversity confronting nation States, particularly oil producing developing states has compromised their zeal to tackle crude oil induced market and environmental problems - stockpiling crude.

1. Introduction

Crude oil which is one the variants of fossil fuel is the most important energy source and the largest traded commodity in the global market. Its derivatives including gasoline, diesel kerosene, jet fuel, asphalt and other petrochemicals utilized in plastic, insecticides, fertilizers etc provide a platform for socio-economic development in most oil producing and consumer nations. This is because of their utilization for cooking, heating homes aviation fuel for operating aircrafts, automobiles, machines, tarring roads, increasing soil fertility, cosmetics and trade commodities. This accounts for volatility of crude oil determining the inflationary trend in food, housing, transportation, investment and services.

The utility of crude oil as a tool for growth has been shrouded by the cartelization of crude oil production by the organization of Petroleum Exporting Countries (OPEC) with increasing tendency to utilize oil as a tool for political pressure. Its non renewable nature and the environment cost of crude oil production have led to increased clamour for its replacement as a primary energy source. the discovery and exploitation of shale oil in commercial quantities by the United States of America and the attendant euphoria generated by the anticipated shift in crude oil demand to oil shale is yet to have an appreciably impact on the crude oil hegemony. Although it has been touted as an energy revolution oil shale discovery has reenacted crude oil pricing as the barometer for global economic trend as oil shale is proving to be more of an energy evolution than a revolution.

2. Crude Oil Production Process

Crude oil production refers to the activities related to surveying drilling, appraisal, development and production of oil and gas, decommissioning and reliabilitation process. These processes shall be analysed as follows:

a) Exploration Surveying: This involves the search for hydrocarbon bearing rock formation to identify productive landscape formation. Dynamite is often utilized as an energy source to access crude deposit.
b) Exploration drilling: On the discovery of a viable hydrocarbon reservoir, an exploratory borehole or well is drilled to conduct preliminary testing. Drilling rigs the capacity to drill through thousands of metres of the Earths crust. Acidization which involves pumping acid or sand into oil reservoirs to facilitate oil or gas extraction appraisal well. On the successful completion of the preliminary drill, more wells are dug to assess the nature size and extent of work required to exploit the crude oil reserve and production rate of a field.
c) Crude Oil Development and Production: Using appraisal wells as indicator, subsequent wells referred to as production or development wells are dug depending on the size of the oil field. Such oil wells range from one to hundreds. Heavy drilling machines are utilized. Gas and water chemicals or heat is usually pumped into the reservoir to sustain the requisite pressure to improve the efficiency of recovery crude oil. Hydraulic fracturing utilized to boost crude oil production results in surface spills and run off during drilling and completion process waste disposal.
d) Decommissioning: This involves removal of structures and equipment utilized in oil exploration and production. Waste disposal is an essential component of decommissioning. Disposal of waste at sea is a common option exercised by oil companies with the ever present threat of contamination[1].

3. Environmental Impact of Crude Oil Production

Crude oil like other fossil fuel is a carbon based energy source which as the global primary energy provider has environmental implications that are far reaching, transboundary and could be severe in its adverse environmental impact depending on the level of unsustainability in its exploitation.

The burning of crude oil by humans as petrol for automobiles and machines, kerosene, diesel, jet fuel heating of homes, industrial energy sources

[1] Mark Saunders and Michele Janes "NERC Report on Decommissioning offshore structures – First Report 1996" (1996) 100 GLTR 430.

and petrochemicals by developed and developing countries has led to atmospheric build up of green house gases (GHGs)[2] which is the major contributor to global warming and climate change.[3] Climate change impacts on physical and biological systems, hydrology and glacial retreat. Some social and economic systems have also been affected by increasing frequency of droughts and floods.[4] Natural ecosystems are particularly vulnerable to climate change due to their limited adaptive capacity which could result in significant irreversible damage to glaciers, coral reefs, mangroves prairie wetlands, topical forests and grassland[5] as well as seasonal changes.

In Nigeria, gas associated with crude oil during exploration process is flared.[6] This has resulted in significant interference with atmospheric carbon balance occasioning acid rain and health challenges including asthma, cancer, skin irritation and premature births and death for inhabitants of the Niger Delta, Nigeria's oil producing region.[7]

Due to years of agitation and government apparent insensitivity to their plight, pipeline vandalisation, ethnic militancy[8], bombing of oil installation[9],

---

[2] George (Rock) Pring, Alexandra Suzann Haas, and Benton Tyler Drinkwine "The Impact of Energy on Health, Environment and sustainable Development: The TANSTAAFL Problem" in Beyond the Carbon Economy eds Donald N. Zillman, Catherine Regwell Yinka. O. Omorogbe, Lila Barrera Hernandez (Oxford, New York: Oxford University press, 2010) p.15 in the United States of America, the producer of 25 percent of the global carbon dioxide emission 88 percent of its human induced GHGs is traceable to its 85 percent fossil fuel reliance.

[3] Inter governmental panel on climate change (IPCC) 2007 Synthesis Report: Summary for Policy makers (Cambridge: Cambridge University Press, 2007) p. 30

[4] James McCarthy Climate Change 2001: Impacts, Adaptation and Vulnerability Contribution of Working of Working Group 11 to the Third Assessment Report of the Intergovernmental Panel on Climate Change (Cambridge: Cambridge University Press 2001) p. 11

[5] Ibid.

[6] UNDP/ World Bank Energy Sector Management Assistance Programme (ESMAP) Strategic Gas Plan (Nigeria) February 2004 confirmed that Nigeria has been flaring over 70 million cubic metres of gas daily for over 40 years

[7] Ibid

[8] In Nigeria, due to the unsustainable exploitation of crude oil and the environmental degradation in the Niger Delta ethnic militia masquerading as environmental defenders like Movement for the Emancipation of the Niger Delta (MEND) Niger Delta People Volunteer Force (NDPVF), Movement for the Emancipation of Ogoni People (MOSOP) emerged to agitate for increased participation of the people in the sharing of oil revenue, infrastructural development and greater environmental protection for oil producing communities.

[9] In May 2008, the Shell Petroleum Development Company (SPDC) Diebu creek flow station was blown up by militants operating along the creeks of southern Ijaw local government area of Bayelsa State causing massive spills. MEND attacked a Shell Nigeria Exploration and Production Company (SNEPCO) storage and offloading vessel which led to shut down of 225,000 barrels per say of crude oil output. MEND also blew up the Opobo pipeline in the Niger Delta while the Martyrs Brigade caused a huge pipeline blaze in a dynamite attack that led to the death of eight persons. See Violet Aigbokhaevbo "Environmental Terrorism in the

oil bunkering[10] and sabotage are often adopted by host communities to protest their impoverishment by unsustainable exploitation of crude oil in their locality. The environment is often the victim of such protest.

Oil drilling is flawed with environmental and health risks Acid, chemicals and dynamite utilized during process are hazardous to the environment. Improper disposal of fluids and chemicals utilized in the process contaminate underground and surface water. Oil well blow –outs is also a complication that could occur in the exploration process.[11]

Oil spills in the course of crude oil recovery, corrosion of pipeline and equipment failure compound environmental challenges.

Off shore oil exploration using seismic technology adversely impacts on the marine environment. It can lead to hearing impairment in whales, fishes and other aquatic animals. The migratory pattern and communication system is also disrupted[12] its effect on shore, range from wildlife disruption, road construction which often interferes with natural habitats to contamination of water sources and land[13]

In an effort to regulate oil bunkering and illegal crude oil exploitation, in developing countries like Nigeria, stolen crude oil and the tankers or vessels and equipments utilized in the theft are confiscated and set ablaze. This further pollutes the air, land and water.[14]

Crude oil production process and the consequential environmental pressure has led to continuous conflict between host communities and

---

Niger Delta: Implications for Nigeria's Developing Economy" in International Energy Law Review 2 (2009): 41.

[10] Oil bunkering is an endemic problem in Niger Delta inspite of effort by government to curb the menace. Nigeria loses approximately 200,000 barrels a day representing about 10 percent of the country's production to pipeline vandals. Leakages from the process contaminate the soil, surface and underground water sources, see Jon Gambrell "Oil Bunkering Threatens Nigeria's Economy, Environment" National, July 21, 2013

[11] Aglionby J. "Oil Drilling Blamed For Fatal Java Mudflow London Financial Times, 8 March 2007, blowout of an oil well in Indonesia released volcanic mud of approximately 200,000 metres daily leaving 15, 000 people homeless and killing and destroying properties in the process.

[12] George (Rock) Pring, Alexandra Suzann Haas and Benton Tyler Drinkwine "The impact of Energy on Health, Environment and sustainable Development: The TANSTAAFL Problem" in Beyond the Carbon Economy eds. Donald N. Zillman Catherine Redgwell , Yinka Omorogbe (Oxford: Oxford University Press, 2010 ,

[13] Ibid

[14] Samuel Oyadongha "Crude Oil theft : JTF seizes three vessels, destroys 363 illegal refineries , Samuel Oyadongha Vanguard January 27, 2013. In Nigeria, the joint task forces known as "Operation Pulo Shield" have been launching operation along the creeks and water ways of Delta and Bayelsa States leading to the destruction and burning of seized vessels and hundreds of illegal refineries.

multination corporations engaged in the oil sector due to the impoverishment of residents due to extensive contamination of rivers and farmlands which make them bad for fishing and farming has made them resentful of multination corporations and the government.

According to the United Nations Human Development Report (UNHDR) on the Niger Delta

> ... the Delta today is a place of frustrated expectations and deep rooted mistrust. Unprecedented restiveness at times erupts in violent. Long years of neglect and conflict have fostered siege mentality especially among youths who feel they are condemned to a future without hope and see conflict as a strategy to escape deprivation.[15]

This has led to increasing sabotage of oil exploration[16] with environmental consequences.[17] The thousands of kilometer of pipeline layout which are sometimes cross border[18] makes adequate policy against vandals and saboteurs impracticable. Energy security and cross border pipeline network are inter related.[19] Russia and Georgia are important transit countries for cross border pipe line the growing tendency of oil and gas being far away from their consumption point has increased inter connecting pipeline the Trans. Afghan pipeline and Nabuco pipeline intended to transport oil from the Caspian region to Europe and from western Kazakhstan to china are examples cross border pipeline network[20]

The transboundary effect of any crude oil production mishap and the challenge of clean up operation which do not completely remove the effect of the spill but continue to compromise the environment several years after the spill renders crude oil production an environmental risk[21] which requires huge capital out lay and technical expertise to tackle effectively.

---

[15] UNDP Niger Delta Human Development Report 2006.

[16] Hamisu Muhammed "Nigeria: Shell Declares Force Majeure on Bonny light" *Daily Trust*, March 6, 2013. Due to oil pipeline spills and rising incidents of oil theft, Shell Petroleum Trunk Line (NCTL) at Nembe Creek had to be terminated. AGIP also had to shut its operation in Bayelsa State due to intensified bunkering with an estimated loss of 7,000 barrels of crude oil per day.

[17] Environmental damage and threat to lives and property resulted in the burning of 52 houses and the death of 7 residents in Bayelsa State see *The Tide Newspaper*, Friday, April 19, 2013.

[18] Ishrak Ahmed Siddiky "The Caspian Energy Scenario and its pipelines: Amalgamation of interests? 1 EIR (2009): 38

[19] Ibid

[20] Ibid

[21] On March 24, 1989 the Exxon Valdez oil spill in the south coast of Alaska Killed hundreds of thousands of birds, fishes and other sea animals. It is considered one of the worst human

4. Global Market Impact of Crude Oil Production

The global market consists of several participants including producer, buyers, transporters, refiners, distributors and consumers whose socio-economic survival are either directly or indirectly relative to the price of crude oil as crude oil is the barometer for determining the global market pricing of goods and services.

The global energy value chain and investment pattern of multinational corporation is often shaped by crude oil pricing. Crude oil is traded on a number of international exchanges for immediate and future delivery. The New York Mercantile Exchange (ICE) in London are the main exchange for trading crude oil. The Brent crude price is the most important benchmark in Europe and is utilized in two thirds of oil trade globally.[22] Soaring oil prices affect the sales and purchase price of goods and services and adversely impact on economic growth.[23] This is because the price of a barrel of crude oil affects the pump price of diesel, petrol kerosene and other crude oil derivative[24]

High oil prices or volatility also influence production storage and consumption patterns which influence efficient resource allocation in the economy. This adversely impacts on the socio-economic development of the country due to imbalance in resource allocation to enable it to shore up the shock arising from its exposure.

Crude oil has metamorphosed into a financial instrument utilized by countries to hedge against inflationary trends or for transferring risk from hedgers to speculators.[25] Prolonged disruption of the crude oil supply chain

---

induced environmental catastrophes. The effect of the spill is still apparent inspite of the clean up efforts as most native species are extinct similarly, on April 20, 2010 the British Petroleum (BP) oil spill due to explosion on its drilling platform caused the oil rig to sink spilling millions of gallons of oil into the gulf of Mexico and adversely impacting on the aquatic environment and humans. see Bourne Joel " Is another deep water disaster inevitable "National Geographic October 2010 p.2

[22] RAC Foundation Deloitte, Uk fuel market review, crude oil www.racfoundation.org/uk-fuel-market review (accessed on March 07, 2013).

[23] Bhanumurthy, Surait Das, Sunkanya Bose "Oil Price Shock Pass through Policy and Its Impact on India" National Institute of Public Finance and Policy Working Paper No 2012 – 99 March 2012 p.3 Rise in global crude slows down economic activities, leads to higher bills, worsens terms of trade, enhances deterioration of trade imbalances and squeeze in aggregate demand for goods and services.

[24] Fleming J, Ostdiek B "The Impact of Energy derivatives on the Crude Oil Market" Energy economics 21(1999) : 136

[25] Ibid

has the potential of destabilizing the global market.[26] Prolonged strike between July and November 2013 in the oil sector in Libya which removed more than one million barrels per day (bbl/d) of crude oil from the global market disrupted the Brent crude oil price path and increased price by more than nine dollars per barrel.[27] Similar disruptions in Nigeria's Niger Delta oil producing region in 2003 resulted in the loss of approximately 800 barrels of crude oil per day. In 2008 Bonga responsible for about 200,000 barrels per day, about 10 percent of Nigeria's crude oil output[28] had to be shut due to host communities sponsored terrorism. Consequently, global oil prices rose by more than 1.5 dollars and Shell BP had to lay off several of its workers.[29] In recent times, rising political tensions in Ukraine as at March 3, 2014 has resulted in higher crude oil prices with the Benchmark West Texas Crude 105.22 dollars and Brent crude oil 111.20 dollars higher.[30] It remains to be seen how significant the market impact of soaring crude prices would be should the Ukraine crisis not be resolved timeously.

5. Quest for Crude Oil Replacement

The realization that global crude oil dependence constitutes a threat to food, energy and political security, due to the disruptive market impact of crude price volatility, has provided the impetus for the quest to replace it as a primary energy source.

The cartelization of crude oil by Organisation of Petroleum Exporting Countries (OPEC) and the possession of 66 percent of the 80 percent world's crude reserve by OPEC member countries[31] has placed immense pressure on global energy security. This is due to the incessant political instability in the Middle East countries which comprise half of OPEC membership and the constant disruption of oil supplies and the threat of sabotage of supplies to developed countries highly dependent on crude oil especially the United States of America. Due to United States tough stance on terror, Islamic

[26] United States Energy Information Administration (USEIA) "Libyan Crude Oil Production Levels Influence International Crude Oil Markets" November 26, 2013.

[27] Ibid.

[28] Giacomo Luciani Armed Conflicts and Security of Oil and Gas Supplies, CEPS Working Document No: 352, June 2011

[29] Hassan Tai Ejibunu "Nigeria's Niger Delta Crisis: Root Causes of Peacelessness" European University Centre for Peace Studies State School training Austria 2007

[30] Gary Strauss "Crisis Ukraine Sends Oil to 2014 highs" *USA TODAY*, March 3, 2014

[31] OPEC share of the World Oil Reserve 2011, OPEC Annual Statistical Bulletin 2012

fundamentalist are increasing the lethality with which they operate. The possibility of terrorist reverting to resource as target terrorism and disrupting oil supplies to the United States and other anti terror countries is a real and ever present danger.[32] Breaking away from the crude oil hegemony has become an imminent necessity for the United States of America.

According to Frank Denton

> ... because we have chosen to rely on Middle East Oil exporters to produce the energy materials which our economy must have, we have ceded to them a power to damage us severely. By maintaining a just enough supply largely with Saudi Arabia leadership, the oil exporters have been able to establish a market for oil that is highly unstable. Whether we act or they act, any event that might bring about reduction in oil supply produces an immediate market response as a premium, around $25 to $35 a barell immediately is added to the resource base of the jihadist[33]

The non renewable status of crude oil and the phobia that over consumption of oil[34] could lead to its depletion has increased the necessity for its displacement as a primary energy source.

The availability of oil shale (tight crude) in commercial quantities in the United States of America, Russia, Australia, Mexico, Argentina and India has encouraged its exploitation as a means of attaining energy security and reducing dependence on OPEC crude supplies.[35]

The United States of America has the global largest shale deposit and its domestic shale deposit and its exploitation is estimated to rise up to 5 million barrels per day by 2017.[36] Other countries possess an estimated 790 billion cubic meters of oil shale deposit out of which 160 billion cubic meters are technically recoverable are at various stages of taking advantage of their oil shale deposit[37] to complement their crude oil energy source.

---

[32] Frank Denton, Nexus Oil and Al Queda, American Energy Independence March 2007 www.AmericanEnergy independence.com /nexus.aspx (accessed on January 25, 2013)

[33] Ibid

[34] International Energy Agency (IEA) Oil Market Report July 11, 2013 estimated that global oil demand to grow by 1.2 mb/d in 2014 due to upwardly revised growth of 930 kb/d in 2013.

[35] Oil Shale resource s is available in the Green River formation Colorado, North Western Utah and South western Wyoming contain more than eight trillion barrels of oil shale. See Anthony Andrews "Oil Shale: History, incentive and Policy" CRS Report for congress order code RL 333 59, April 13, 2006.

[36] Maugeri Leonardo "The Shale Oil Boom: A U.S Phenomenon" Discussion Paper 2013 – 05 Belfer Centre for Science and International Affairs, Harvard Kennedy School, June 2013

[37] International Energy Agency, World Energy Outlook, 2010.

Central and Eastern European (CEE) countries are however adopting a cautious approach to oil shale exploitation due to concerns about the environmental hazardous consequences of its exploitation process.[38]

In spite of this, oil shale is rapidly emerging as a significant challenge to crude oil dominance. It is anticipated that it could revolutionize global energy market by providing long term energy security at lower prices for several counties.[39]

It has however been argued that the global economy has been crude oil dependant for so long that due to the huge resources outlay and the drilling intense nature of the oil shale recovery process and the absence of private mineral rights in most countries, the oil shale evolution may not be the energy revolution it has been touted to be.[40]

The risk of earth tremors[41] or minor earth quakes renders oil shale exploitation an environmentally unattractive replacement for crude oil. Its hydraulic fracturing recovery process and the chemicals utilized contaminate surface and underground water while the large volumes of water required for efficient oil shale recovery poses significant threat to water security in host communities.[42]

Until a more efficient cheaper and cleaner energy source is discovered, crude oil remains the determinant of global market performance and the environmental impact of its exploitation, storage, transportation and utilization is a continuous challenge that has to be addressed by present and future generations of humans.

6. Recommendations

Host communities should be allowed to benefit more from their crude oil resource exploitation through payment to them of adequate royalties by

---

[38] Paul Stevens "The Shale Gas Revolution: Development and Changes" Chatam House Briefing Paper, Energy, Environment and Resources EERBP 2012/04 August, 2012.

[39] PWC Shale Oil: The next energy revolution February 2013 www.pwc.co.uk (accessed on August 1, 2013)

[40] Peter Clover and Michael Economides (New York: The Continuum International Publishing Group, 2010)

[41] Leonardo Maugeri "Oil: The Next Revolution Discussion Paper #2010 – 12 Geopolitical Energy project, Belfer Centre for Science and International Affairs, Havard Kennedy School, June 2012

[42] KPMG Global Energy Institute, Central and Eastern Europe Shale Gas Outlook 2012.

government[43] and more efficient implementation of corporate social responsibility (CSR) by multi-national corporation. This is to reduce incidents of pipeline vandalisation, oil bunkering and sabotage which often add to the financial and environmental cost of exploiting crude oil.

Concerted global effort to promote the utilization of cleaner and renewable energy sources should be sustained. This is through the systematic phasing out of crude oil and its derivatives dependent technologies and replacing them with more accessible and affordable solar, wind or hydro powered equipments. Should this be impracticable, on the short run, crude oil utilization should be reduced to the barest minimum through the development of energy efficient technologies

Sustainable exploitation of crude oil should be strictly enforced through greater oversight management of regulatory agencies to ensure proper balancing of environmental concerns and profit generation. This is to reduce the incidents and magnitude of environmental crimes.

To reduce the market impact volatility of crude oil, countries should be encouraged to exploit their domestic alternative energy resources. This would reduce their crude oil dependence and significantly dilute the ability of crude oil to dictate the market trend.

Due to the transboundary effect of crude oil pollution and the ability of its deleterious impact to continue for several years after the spill, exploiters, transporters and users of crude oil should be mandated to employ ultimate care in handling the product by taking all practicable precautions and using up to date equipment to ensure that oil pollution is contained when it occurs.

Public and operator enlightenment on the environmental impact of crude oil exploitation and utilization should be sustained to encourage environmental friendly mode of oil consumption. Regular public enlightenment is an essential component of inculcating environmental sensitivity in the populace.

The polluter pays principle should be strictly enforced in the oil sector to render oil companies more accountable for their polluting activities. In case of consistent violation and anti pollution, regulation the operator's licence should be revoked as a deterrent to would be offenders.

Oil shale and other alternative energy source should be encouraged to diversify the global energy mix while stressing the necessity for environmental

[43] Section 162(2) of the 1999 amended constitution provides for the allocation of not less than 13 percent of revenue accruing to the Federation account to the state from which the natural resource was derived see the supreme court reiteration of this position in A.G Federation v A.G Abia State (No 2) (2002) 6NWLR (pt 764) p. 542 particularly p.636

sensitive mode of exploitation. Apart from reducing the market impact of crude oil, it would enhance global energy security.

Environmental impact of crude oil can only be minimized and not eliminated as pollution is a natural consequence of crude exploitation process. It is recommended that reduced consumption of crude oil derivatives is the most effective means of reducing its environmental hazardous effect.

CONCLUSION

Crude oil dominance as a global energy source constitutes the greatest threat to energy security, market stability and environmental intergrity. The global trend of stockpiling crude oil as a market volatility check has to be reviewed to encourage Nation States to instead make concerted effort to exploit renewable alternative energy sources. Unless this is done crude oil will continue to dictate global environmental health and socio-economic development to the frustration of most nation states.

In: Crude Oils
Editor: Claire Valenti

ISBN: 978-1-63117-950-1

Chapter 6

A FASCINATING CHALLENGE FOR PETROLEUM INDUSTRY: NEW MODELS FOR PREDICTING FUEL ECONOMY PERFORMANCE OF LUBRICANT OILS USED IN AUTOMOTIVE ENGINES

José-Alberto Maroto-Centeno [*,1], ***Tomás Pérez-Gutiérrez*** [†,2], ***Luis Fernández-Ruíz-Morón,*** [‡2] ***and Manuel Quesada-Pérez*** [§,1]

[a]Group of Physics and Chemistry of Linares, Department of Physics, EPS of Linares, University of Jaén, Linares, Jaén, Spain
[b]Technology Unit., Repsol, Móstoles, Madrid, Spain

ABSTRACT

Due to increasingly strict vehicle fuel economy mandates over the past two decades, fuel economy improvement continues to be a focal point in all aspects of engine and vehicle engine and operation. This

* corresponding author: e-mail: jamaroto@ujaen.es, Tel.: +34 953 648 553, Group of Physics and Chemistry of Linares, Departament of Physics, EPS of Linares (University of Jaén), C/ Alfonso X el Sabio, 28, 23700 Linares, Jaén, Spain
† e-mail: tperezg@repsol.com
‡ e-mail: lfernandezr@repsol.com
§ e-mail: mquesada@ujaen.es

includes engine oil formulation, whose fuel economy improvement potential can be estimated in the interval from 1% to 4%, depending on the chosen baseline. In response, all major global regions have established standard engine oil fuel economy tests, whether through industry groups (such as ILSAC, API, or ACEA), or through individual OEMs. In addition to the standardized fuel economy tests that oils are required to pass to meet specifications, bench tests have historically been used to screen and assess the fuel economy performance of these oils. It is generally accepted for engine oils that fuel economy improvement is influenced by reductions in kinematic viscosity, high shear viscosity, boundary friction, thin-film friction and pressure-viscosity coefficient (or traction coefficient in elastohydrodynamic lubrication). These measurements are relatively quick and easy to obtain compared to the more sophisticated and expensive engine performance tests. Therefore, optimization of bench tests can be considered as a fascinating challenge for petroleum industry. Nevertheless, in order to evaluate the fuel economy performance of lubricant oils by means of bench tests it is required the previous design of a precise predictive model. Currently, the Spanish company REPSOL S.A. in cooperation with the Research Group of Physics and Chemistry of Linares (University of Jaén-Spain) is developing a new model to characterize the CEC L-54-96 standard engine fuel economy test which is part of the standards ACEA A1/B1, A5/B5, C1, C2, C3 and C4. Preliminary results and reflections are shown in this commentary.

Due to increasingly strict vehicle fuel economy mandates over the past two decades, fuel economy improvement continues to be a focal point in all aspects of engine and vehicle engine and operation. This includes engine oil formulation, whose fuel economy improvement potential can be estimated in the interval from 1% to 4%, depending on the chosen baseline [1]. In response, all major global regions have established standard engine oil fuel economy tests, whether through industry groups (such as ILSAC, API, or ACEA), or through individual OEMs. Standardized fuel economy tests of lubricants are either in common use in the USA, Japan and Europe. Example of such tests are the ASTM Sequence VID and the CEC L-54-96 standard engine fuel economy test which is part of the standards ACEA A1/B1, A5/B5, C1, C2, C3 and C4. These tests show that carefully formulated lubricants can make significant contributions in order to reduce gasoline consumption by the optimization of internal engine friction. Nevertheless, the selection of optimal lubricant rheology and surface chemical properties to yield high fuel efficiency is quite complex because the overall friction within an operating engine originates from several different engine components, including the valve train, piston pack and bearings [2]. These components each subject the lubricant to different and widely-varying conditions of temperature, load and shear rate throughout an engine cycle. The net result is that there exists, within a firing engine, a balance of the different regimes of lubrication: hydrodynamic, boundary and elastohydrodynamic (EHD) lubrication.

In addition to the standardized fuel economy tests that oils are required to pass to meet specifications, bench tests have historically been used to screen and assess the fuel economy performance of these oils. These measurements are relatively quick and easy to obtain compared to the more sophisticated and expensive engine performance tests. Therefore, optimization of bench tests can be considered as a fascinating challenge for petroleum industry. Nevertheless, in order to evaluate the fuel economy performance of lubricant oils by means of bench tests it is required the previous design of a precise predictive model. A plausible possibility [2-4] is the use of the high shear viscosity (η_{HTHS}) to characterize the hydrodynamic lubrication regime, the use of the boundary friction coefficient (μ_B) to characterize the boundary lubrication regime and the use of the EHD traction coefficient (μ_{EHD}) to characterize the elastohydrodynamic lubrication regime. Also, we decide to use the following equation that assumes a linear correlation between the fuel economy increment (*FEI*) and each of these three parameters:

$$FEI = A' + B'\eta_{HTHS} + C'\mu_B + D'\mu_{EHD} \tag{1}$$

where A', B', C' and D' are adjustable parameters.

It must be underlined that linear correlations between *FEI* and η_{HTHS} and between *FEI* and μ_B have been experimentally proved [5-7].

Currently, the Spanish company REPSOL S.A. in cooperation with the Research Group of Physics and Chemistry of Linares (University of Jaén-Spain) is developing a new model to characterize the CEC L-54-96 standard engine fuel economy test which is based in the use of this mathematical expression. The experimental measurements are being carried out in the Repsol Advanced Laboratory of Lubricants, which is placed in the Repsol Technology Centre at Móstoles (Madrid, Spain). The key parameters of 22 lubricants oils are being evaluated in order to elucidate the value of the adjustable parameters A', B', C' and D'. Concretely, the high shear viscosity (η_{HTHS}) is being measured at 150 °C by means of a Ravenfield Viscometer (Method ASTM D 4741), the boundary friction coefficient (μ_B) is being measured at 100ºC by means of a High Frecuency Reciprocating Rig (HFRR) and the EHD traction coefficient (μ_{EHD}) is being measured at 100ºC and at a velocity of 2500 mm/s by means of a Mini Traction Machine (MTM) in a configuration ball-on-disk. Measurements of *FEI* are being supported by an external company. After both preliminary measurements and multiple linear regression analyses have permitted us a first evaluation of parameters *A´*, *B´*, *C´*and *D´* and the drawing of the following sectors diagram which illustrates the role played by the three parameters evaluated (η_{HTHS}, μ_B and μ_{EHD}) in the fuel economy behavior of a lubricant oil which is tested by means of the CEC L-54-96 standard engine fuel economy test.

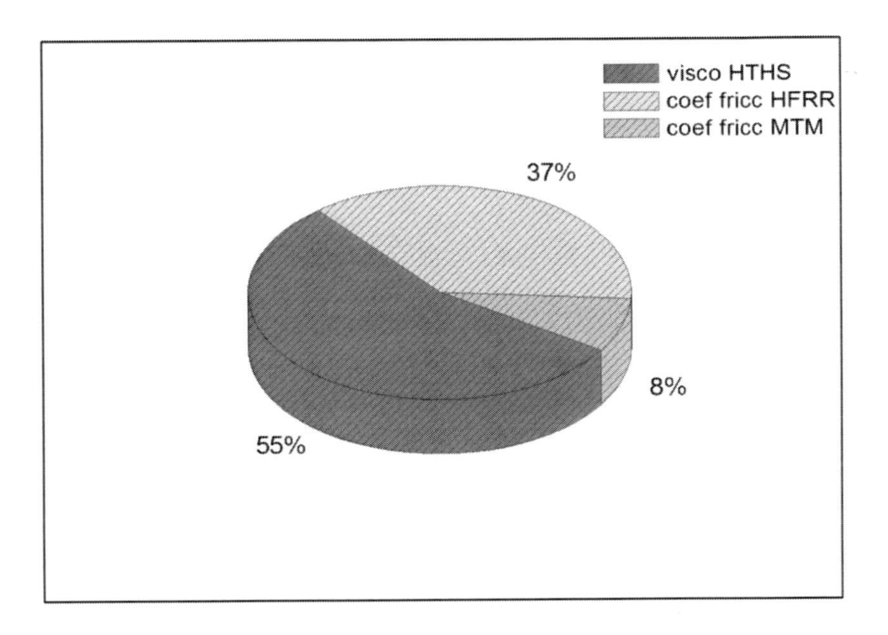

In the light of this sectors diagram it is evident that the high shear viscosity is going to play a fundamental role in the fuel economy behavior of a lubricant oil and, as consequence, the use of a suitable Viscosity Index Improver will be highly advised in order to formulate new fuel economy lubricant oils. On the other hand, the use of friction modifiers can help to optimize the friction in the boundary lubrication regime. Finally, the sector diagram shows that the role played by the elastohydrodynamic lubrication regime in the fuel economy behavior of a lubricant oil is low but not negligible. The use of equation (1) will permit us to quantify the exact influence of these additives in the *FEI*. As conclusion, the simultaneous use of equation (1) and the sectors diagram are useful tools that will facilitates the formulation of new fuel economy lubricants. Also, important economic saving can be expected by means of the use of laboratory bench tests in a predictive way because the extensive use of standardized fuel economy tests for direct evaluation of *FEI* are very expensive [1].

REFERENCES

[1] J. Styer and G. Guinther, "Fuel Economy Beyond ILSAC GF-5: Correlation of Modern Engine Oil Tests to Real World Performance", *SAE Int. J. Fuels Lubr.* 5(3), 1025, 2012.

[2] C. Bovington, V. Anghel and H.A. Spikes, "Predicting Sequence VI and VIA Fuel Economy from Laboratory Bench Tests", *SAE paper* 961142, 1996.

[3] J. Igarashi, M. Kagaya, T. Satoth and T. Nagashima, "High Viscosity Index Petroleum Base Stocks – The High Potential Base Stocks for Fuel Economy Automotive Lubricants", *SAE paper 920659*, 1992.

[4] M.T. Devlin, W.Y. Lam and T.F. McDonnel, "Critical Oil Physical Properties that control the Fuel Economy Performance of General Motors Vehicles", *SAE paper 982503*, 1998.

[5] D.W. Morecroft, "The Shell low-velocity Friction Machine for Evaluating Fuel Economy Motor Oils", *Wear*, 89(29, 215, 1983.

[6] G.R. Dobson and W.C. Pike, *"Predicting Viscosity Related Performance of Engine Oils"*, Erdol und Köhle, 36(5), 218, 1983.

[7] G.R. Dobson, "The Prediction of Fuel Efficiency of Engine Oils", *CEC Symposium on the Performance of Alternative Fuels and Lubricants,* 1981.

INDEX

D

E

F

G

H

I

J

K

L

M

N

O

P

Q

R

S